HEAD
SHOT

HEAD
SHOT

THE SCIENCE BEHIND
THE JFK ASSASSINATION

G. PAUL CHAMBERS

Prometheus Books

59 John Glenn Drive
Amherst, New York 14228–2119

Published 2010 by Prometheus Books

Inquiries should be addressed to
Prometheus Books
59 John Glenn Drive
Amherst, New York 14228–2119
VOICE: 716–691–0133
FAX: 716–691–0137
WWW.PROMETHEUSBOOKS.COM

14 13 12 11 10   5 4 3 2 1

Library of Congress Cataloging-in-Publication Data

Chambers, G. Paul, 1957–
    Head shot : the science behind the JFK assassination / by G. Paul Chambers.
        p. cm.
    Includes index.
    ISBN 978–1–61614–209–4 (cloth : alk. paper)
    1. Kennedy, John F. (John Fitzgerald), 1917–1963—Assassination. 2. Forensic ballistics—United States—Case studies. I. Title.

E842.9 .C46 2010
363.25'.62—dc22

                                                                        2010020029

Printed in the United States of America on acid-free paper

To Rachel and Alexis

CONTENTS

INTRODUCTION

The assassination of JFK was the crime of the century. It has spawned two generations of literature and a lifetime of debate. Why another book on the assassination? Why now?

In September 1964, the Warren Commission published its final conclusions on the death of John F. Kennedy. Since that time, the report has been a target of criticism. Their conclusion that Lee Harvey Oswald killed Kennedy, and acted alone, has been challenged by numerous authors over the last forty years, and by the establishment of a series of government committees to reinvestigate the assassination. Most polls show that a majority of Americans do not believe that Oswald acted alone. Clearly, the public has never been satisfied by the Warren Commission report. Yet the conclusions of this report are still pronounced valid today by not only our government, but also by the mainstream press.

The percentage of Dallas residents who believe that a conspiracy was behind Kennedy's death approaches something like 90 percent. The day after the assassination, the *Dallas Morning News* of November 23, 1963, carried a story stating:

> [District Attorney Henry] Wade said preliminary reports indicated more than one person was involved in the shooting which brought death to the president and left Gov. John Connally wounded...." This is the most dastardly act I've ever heard about," Wade said. "Everyone who participated in this crime—anyone who helped plan it or furnished a weapon, knowing the purposes for which it was intended—

is guilty of murder under Texas law. They should all go to the electric chair."

However, after several phone calls from President Johnson's aide Cliff Carter, Wade abruptly changed his story.[1] All of Wade's subsequent statements focused on Oswald. To entertain the possibility of a conspiracy opened the possibility that a Soviet plot was behind Kennedy's death. This linkage would at minimum have disrupted the delicate cold war relations between the United States and Russia.[2] It was therefore essential that Washington "manage" the news as much as possible during the early days after the assassination.

Since then almost a thousand books have been published on the Kennedy assassination; most have uncovered serious flaws in the official version of events, but some are, in my view, irresponsible and play into improbable and impossible themes. Among these are certain critical and influential books such as Gerald Posner's *Case Closed* and Vincent Bugliosi's *Reclaiming History*, which, I would argue, paint a distorted picture of the truth. Many in the news media and in the general public now consider Vincent Bugliosi's *Reclaiming History* the last word on the subject. However, his physics is wrong, and his science is, frankly, impossible. Although Bugliosi's extensive work is hopelessly flawed, it has convinced many people that the government actually got the facts right the first time, that no conspiracy ever existed, and that Kennedy's death was the work of a single "lone nut," which masks the truth of the assassination. A realistic scientific treatment of the assassination is sorely needed now to clear up the misinformation disseminated by these popular and prominent works on the subject.

As the forty-seventh anniversary of the assassination of President John F. Kennedy approaches in November, I write this book to critically and accurately examine the science and the physics behind the assassination. Based on my experience in physics, this volume will scientifically resolve one of the enduring mysteries of the twentieth century. The approach I take draws on my fifteen years as a research physicist for the US Navy. My background in detonation physics and high-speed

INTRODUCTION

photography of explosive events allows me to properly and accurately assess the scientific data set associated with the assassination.

I will show from careful scientific analysis that the final, fatal bullet that struck Kennedy was not fired from the Mannlicher-Carcano rifle purported to have fired all the shots in Dealey Plaza that day. Although many pro-conspiracy books on the assassination have been written, as yet, no one has identified the weapon that fired the final fatal round. I will do so. I will further identify the precise locations of the assassins. This analysis pulls the rug out from under the conclusions of the Warren Commission, namely, that Oswald acted alone and that no shots were fired from the now infamous grassy knoll.

I am a research scientist. I have no axes to grind. I do not belong to any groups or organizations seeking to prove government conspiracies, advance cultural agendas, or promote some political action. My approach is singularly that of a detached researcher investigating history through application of the scientific method. My treatment is beyond politics and "theories." My goal is to assess the existing empirical evidence and add to it to reach the best possible conclusions.

The first three chapters of this work are devoted to the Warren Commission report and to the testimony of witnesses present at the time of the assassination. The detailed problems with the report and its inconsistencies are highlighted. The fourth chapter is a discussion of the history and the development of the scientific method. It lays a foundation for the rest of the work and outlines the program needed to uncover the truth of the assassination. For those readers with a strong background in science and its history, or for readers who are less interested in how the methods of science developed and more focused on the assassination itself, this chapter could be read with less attention to detail in the hope of more quickly reaching the core chapters relating to the evidence. However, I do not advise such a course of action. We are attempting to unravel one of the most horrible crimes of the twentieth century. The material presented here is highly relevant to resolving the assassination. It is always valuable to reacquaint ourselves with the basic elements of good scientific reasoning. They will serve us well in the chapters ahead.

Chapters 5 through 9 apply this scientific method to find the truth of the assassination. This is the core of the book. The same methods that a research scientist would use to analyze a high-speed event, such as a detonation, are employed here. To the extent that science is believable, one may have confidence in this method of analysis. All one has to do to see real science in action is turn on a computer, a flat-screen television, or a GPS device. These devices provide convincing demonstrations of some of the best science the twentieth century had to offer. The last chapter explains why Kennedy's assassination is still relevant today. To deny Americans the truth about the assassination is literally to deny us our history.

Sadly, the passing of two of the greatest science writers of our generation, Carl Sagan and Stephen Jay Gould, has left a void in the field of science awareness for the general public. Although I could never aspire to be as penetrating or prolific as either of these luminaries, I hope that this humble effort will in some small way contribute to filling this void. I strongly believe that most people have the capacity to know the truth when they hear it. I hope this modest contribution to the literature will strike a chord.

ACKNOWLEDGMENTS

I would like to thank my agent, Ron Goldfarb, for helpful discussions, for pointing me in the right direction, and for taking an amorphous project and turning it into something substantive. I would like to acknowledge G. Robert Blakey for some very helpful and informative discussions. I would also like to thank Ms. Mary Kay Schmidt of the Special Access and FOIA Staff at the National Archives for her expert and patient assistance in finding the assassination records I needed. I would like to thank my wife, Linda Alvey-Chambers, for doing the artwork. I would like to thank my sister, Diane Chambers, for patiently reading the manuscript and providing much-needed corrections. I would like to thank my daughter Rachel Chambers for her kind assistance with the preparation of the index. It goes without saying that any errors are mine alone.

CHAPTER 1

THE WARREN COMMISSION

The single bullet theory is no longer a theory. It's a fact.
—Arlen Specter, former head of the
Senate Judiciary Committee

As President Lyndon B. Johnson winged back to Washington on Air Force One following his taking the oath of office in Dallas, he had one predominant thought on his mind: were the missiles already flying? He could only assume—he could only afford to assume—that the assassination of President John F. Kennedy was the first act in a premeditated attack against the United States, an overt prelude to World War III. With Kennedy suddenly gone, the US government was in temporary confusion and disarray. What better time to launch the all-out attack that seemed all but inevitable since the Cuban Missile Crisis that had been touched off one year earlier and had brought the United States to the brink of nuclear confrontation with the Soviet Union? If Kennedy's death was the end result of a Soviet plot, the prospect of war now seemed virtually imminent.

But even if Kennedy's murder was not the 1960s version of Pearl Harbor, the very real threat of nuclear war still loomed before the nation. Although presidents had been assassinated several times in the past, the specter of the involvement of foreign powers had never before been an issue. The prospects of international war had not arisen. Now, in the nuclear age, the possibility of the involvement of a foreign nation was all too real. The stakes had never been so high.

Johnson was still old enough to recall that World War I, the war to end all wars, was touched off by an assassination. Within a month after the unfortunate and unlucky murder of Archduke Ferdinand of Austria, the whole of Europe was at war.

Archduke Ferdinand was killed by the most unfortunate series of events imaginable. His driver got lost returning from an official visit to City Hall in Sarajevo, Yugoslavia, on a June morning in 1914. Somehow, improbably, he ended up near a tavern where a group of young conspirators were waiting. They had given up on assassinating the archduke, but when he appeared before them, they decided to go through with their plot. The assassin, nineteen-year-old Gavrilo Princip, was consumed with a fierce feeling of Slavic nationalism. He believed the death of the archduke would free his people from the Austro-Hungarian Empire. This act had unexpected political implications.

A third country, Serbia, played a key role in the assassination. Independent Serbia had provided the guns, ammunition, and training for the assassins. The Balkan region erupted in turmoil and open conflict. Within a month, all the major countries in Europe, including Russia, had geared up for war. Every nation thought that every other nation was getting ready to attack and no one wanted to be caught unprepared. As a result, all the major Western powers mobilized and in 1914, mobilization meant war. This terrible tragedy, which impacted and defined the twentieth century, was caused by a singularly unfortunate and ill-fated event.

Lying in the trenches, having survived the war miraculously by his own account, was a young corporal named Adolf Hitler. Hitler never forgot the First World War, never forgot the shame of Germany's defeat, and never forgot how his country was mistreated and humiliated by the Treaty of Versailles when they had not even really lost the war; they had just decided to stop fighting. Germany had retired from the field as a super elite fighting force. Yet Germany was obliged to accept very unfavorable treatment after the war.

This treatment rankled Hitler throughout his lifetime and in due course he found a way to attain revenge by using Germany's disgrace to rise to power. The war he fought in 1914 and its unfavorable outcome were always on his mind. Attacking the world was a way for Germany to recover her honor, avenge the deaths of the previous generation, and turn the appalling treatment at the hands of Europe to her advantage. The entire holocaust of the first half of the twentieth century was therefore ultimately attributable to a single ill-fated assassination. What would be the ruin of the next fifty years? A nuclear exchange among cold war superpowers threatened to kill more people than all the men and women who died in the First and Second World Wars combined. The terrible weapons that had ended the Second World War now threatened to be unleashed on all of humanity.

In a powder keg situation like the one that now confronted the nation, Johnson's worst fear came in the form of the lighted matches of confusion, speculation, and rumor mongering that surrounded Kennedy's death. Johnson knew he needed to move quickly to quell the inevitable hearsay, gossip, and half truths that would have arisen and threaten to spin out of control. With his Executive Order 11130, signed within a week of his oath of office, Johnson set up a commission to "ascertain, evaluate and report upon the facts relating to the assassination of the late President John F. Kennedy." This act not only placed the nation on the path to determine the truth of what happened to Kennedy, but also would put a stop to other independent investigations like those that were getting under way in Dallas. The spectacle of multiple investigations would have been difficult for the public to embrace, would have led to confusion, and would have fanned the flames of the rumor mill that threatened to engulf the nation during the perilous days following Kennedy's death.

The seven-man commission was to be headed by US Supreme Court Chief Justice Earl Warren and included Gerald Ford, representative from Michigan; Allen W. Dulles, former director of the CIA; John J. McCloy, former president of the World Bank; Richard B. Russell, senator from Georgia; John S. Cooper, senator from Kentucky; and Thomas H. Boggs, House majority whip from Louisiana. All were lawyers. Historically, there had been nothing like this commission in modern history. Edward Jay Epstein, a political science student at Cornell University and one of the earliest assassination researchers, felt that the closest historical analogue to the new commission was the Roberts Commission that had been convened to investigate the Pearl Harbor attack.[1]

Johnson had appointed men for this crucial commission who had the highest possible credibility. What better choice to head the newly created commission than the Chief Justice of the United States, Earl Warren? Filling out the committee, he would choose distinguished leaders in the Senate and House of Representatives as well as former heads of government agencies. Since the government enjoyed great trust at that time, any decision from these leaders should effectively quell all doubts surrounding the assassination and defuse the ticking time bomb of the public's confusion and its inconsolable grief.

But Chief Justice Warren wanted no part of the new commission. His first objection was that he already had a full docket of case work before him in the current Supreme Court term. It would be difficult if not impossible for him to devote the time that such a task would require. Second, the constitutional separation of powers militated against a Supreme Court justice serving on a presidential commission. It would conceivably undermine the well-entrenched concept of judicial independence. Third, the work of the commission might draw litigation and force him to disqualify himself from Supreme Court cases stemming from its work.

But after Johnson explained to him the severity of the rumors floating around the assassination, the potential implications of Cuban or Soviet involvement, and the possibility of these leading to war with the Soviet Union, Warren reconsidered. "He went on to tell me that he had just talked to the Defense Secretary Robert McNamara, who had advised him that the first nuclear strike against us might cause the loss of 40 million people. I then said, 'Mr. President, if the situation is that serious, my personal views do not count. I will do it.'"[2]

Senator Russell didn't want to participate either. Russell, a southern Democrat who had been Johnson's mentor when he was in the Senate, despised Earl Warren because of the Court's 1954 ruling in *Brown v. Board of Education*, which had declared public school segregation unconstitutional. Russell didn't like Warren and he didn't trust him. He certainly didn't want to be on a committee that was to be headed by him. But Johnson browbeat Russell until he capitulated. Thus, two of the commission's seven members had to be convinced to accept the task.

To jump-start the commission, President Johnson also requested a report on the assassination from J. Edgar Hoover, who headed the Federal Bureau of Investigation. On December 9, 1963, the bureau produced a 384-page, five-volume report concluding that Lee Harvey Oswald was the sole assassin and that no evidence of a conspiracy existed.[3] One of the first orders of business for the new commission was to evaluate this report.

The commission held its first meeting December 5. The members believed they clearly had a presidential mandate to conduct an independent investigation. They concluded that "the public interest in insuring that the truth was ascertained could not be met by merely accepting the reports or the analyses of Federal or State agencies."[4] Therefore, the commission would conduct its own independent investigation. In order to facilitate this, a general counsel was needed. Earl Warren proposed J. Lee Rankin, a former solicitor general of the United States. The commission unanimously agreed with this choice. The last order of business was to deal with the independent investigation being conducted in Texas. The commission agreed to ask Vincent Waggoner Carr, attorney general for the state of Texas, to postpone the Texas inquiry until the commission had completed its work. Warren wrote to Carr explaining the potential pitfalls of separate investigations and invited him to participate in the commission's investigations, an invitation that Carr accepted. On December 13, Congress passed a joint resolution empowering the commission to subpoena witnesses and grant immunity from prosecution to avoid Fifth Amendment hindrances to discovering the truth. With the assistance of all federal agencies guaranteed by Johnson, the commission now had all the tools it needed to conduct a meaningful investigation.

On December 16, the commission met to discuss the recently disclosed FBI report. Much of the information in the report had already been leaked to the press. J. Edgar Hoover was apparently not happy with the establishment of a separate entity to investigate the assassination and felt that the public could see the pointlessness of this redundant effort if the results from the FBI's own investigations were placed before them.

Even so, most commission members felt that the report lacked depth. The FBI report did not include details about Governor Connally's wounds. Little information was included concerning Jack Ruby, the alleged killer of Lee Harvey Oswald, his associates, movements, and how he got into the basement of the Dallas City Hall garage the morning of November 24, when Oswald was to be transported to the county jail. Warren believed that he would need the original raw materials, the interviews, photographs, affidavits, and recordings to assess the validity of the report. Despite its reliance on government agencies to perform the investigative fieldwork, the Warren Commission would make its own assessments and determinations of the evidence.

However, most commission members already had full-time jobs. Therefore, the tasks of investigating, determining the facts of the assassination, and developing a model of what actually happened was left to junior staff lawyers. These were men in their early thirties who had exceptional law school records and their own practices. Howard Willens, a young Justice Department lawyer, chose most of these attorneys: "We wanted independent lawyers, not government men, who had been at the top of their class and who could work sixteen hours a day."[5]

This approach is analogous to using postdoctoral research associates to conduct scientific research in academia or government. Postdoctoral associates, or postdocs, are scientists who have just obtained their doctoral degrees but have not yet secured a permanent research position. Many academic departments and government research laboratories are built on the backs of the postdocs, who are young and bright, trained in the latest techniques and eager to do original research. They have more energy, enthusiasm, raw ability, and desperation than do their older, established counterparts. They have short windows to achieve recognition, often only a two- to three-year probationary period. During that time, they must publish research papers in scientific journals to secure their reputations. What they lack in experience they make up for in ambition, drive, and determination.

By contrast, senior scientists and professors are often tied up with administrative duties of one kind or another, responsibilities that require extensive experience, budgetary expertise, knowledge of the field, and well-established interpersonal relationships with other scientists and researchers. This situation is exactly analogous to the situation in which Warren Commission members found themselves. They were all credible experienced leaders, but other than Allen Dulles and John McCloy, they were consumed with other full-time duties. Employing junior people, provided they are given proper guidance, can often be the only viable path to success when faced with such a labor-intensive process.

The downside of this approach, however, is that junior people need extensive supervision. I have seen postdocs go off in the wrong direction, waste their time, make crucial mistakes, and end up out of science. One postdoctoral fellow I knew was an incredibly hard worker but ended up missing the scientific boat entirely. He spent all of his time writing operating system software and making electronic circuit boards. These are good skills to have, but they don't have much to do with science. It's best to purchase software and circuit boards out of project funds and commit your time to doing research that is publishable. Most senior scientists can recount numerous postdoc horror stories in which these junior personnel ended up out of science, not because they weren't intelligent, hard workers, but because they missed the forest for the trees. They either didn't receive the proper guidance from their mentors, or they refused to heed the advice and direction of senior people. Even the best junior researchers simply can't be left to their own devices; they need at least some leadership and guidance from more-established scientists to be successful.

The other problem with provisional junior people is that the organization doesn't have the control over them that it does over permanent company employees. Postdocs are temporary appointments, sometimes arranged through third-party organizations. Sometimes these researchers can't even be fired because they have an appointment for a set time period through another organization, like the National Research Counsel, for instance. The postdoc does not get performance reviews. Postdocs would usually like a good recommendation from their adviser so that they can secure a permanent position, typically at another institution. But they might decide to go off on their own, pursue work in other related areas, or leave science altogether. Because there is no strong direct control over a postdoctoral researcher, an organization would shy away from using these individuals for a high-profile project with a tight deadline. The outcome from relying on junior temporary people could be good or bad, depending on luck and circumstances, and the Warren Commission would sink or swim based on the performance of its junior staff.

Initially, Howard Willens had been asked by Nicholas Katzenbach, the deputy attorney general, to act as a liaison between the Justice Department and the commission.[6] Willens had thought the job with the commission would be a part-time duty, but when he realized the extent of the work before the commission, he packed up his office and moved to the commission's offices. Willens would assume a key role, taking over administrative functions, dividing up the work, scheduling, and making requests to other agencies for assistance.

In addition to Willens, the commission's chief counsel, Lee Rankin, needed another assistant and selected Norman Redlich, a New York University law professor, to fill the post. Redlich would ultimately work on specific projects, like the testimony of Lee Harvey Oswald's widow, Marina Oswald, and leave the daily operations of the commission to Willens. In addition to these men, Rankin chose senior counsel for his staff. Most prominent among these were Francis W. H. Adams, a former New York City police commissioner, and Joseph A. Ball, a prominent California trial lawyer. Because most of these men could not simply give up their private practices and devote full time to the investigation, the bulk of the work fell to the junior counsel, appointed by Willens.[7] These men worked as consultants and received seventy-five dollars a day.

The junior counsel, the people who were the key to the success or failure of the commission, consisted of seven men. Redlich chose Melvin A. Eisenberg, a twenty-nine-year-old lawyer from a New York firm who had been first in his class at Harvard; Arlen Specter, a former assistant district attorney and former coeditor of the *Yale Law Journal*; Samuel Stern, a Washington, DC, lawyer and former law clerk to Chief Justice Warren; Burt W. Griffen, also a former assistant United States attorney; David Belin, an Iowa trial lawyer; W. David Slawson, a Denver lawyer; and Wesley J. Liebeler, a former Wall Street attorney. The commission accepted these men based solely on Rankin's endorsement.

In his book *Inquest: The Warren Commission and the Establishment of Truth*, Edward Jay Epstein reported that by December 28, Willens had drafted a memorandum that established the principles of operation for the commission.[8] This memorandum suggested that the overall investigation be divided into five general areas, assigning a senior and junior lawyer to each area to resolve the relevant issues between them. Only the most major issues were to be brought to the attention of the commission. Each team would prepare a chapter of the final report based on its research.

The first topic of inquiry was to be the basic details of the assassination. The second area would be the identity of the assassin. The third would focus on Lee Harvey Oswald's background and possible motives. A fourth subject area would cover Oswald's relationships and possible co-conspirators. The fifth area would investigate the events and circumstances surrounding Oswald's untimely death.

Willens and Rankin then assigned lawyers to each area. The first area was assigned to attorneys Francis W. H. Adams and Arlen Specter. Their job would be to determine the source of the shots that struck Kennedy. Both men had substantial investigative experience. But Adams would turn out to be a no-show. During the entire investigation, he came to Washington only a few times. He was consumed by duties at his own law firm in New York, the ever-present danger of using established working professionals to perform labor-intensive work. Although Rankin had considered asking for Adams's resignation, he was concerned that this would show signs of dissension on the commission.[9]

Therefore, the monumental task of determining the basic facts of the assassination fell, surprisingly, to a single individual—Arlen Specter. Among the commission consultants, Specter was solely responsible for ascertaining the source of the shots, the number of assassins, the precise manner in which the president and Texas governor John Connolly were shot, and the sequence of events. This was a great deal of responsibility and an overwhelming amount of work for one person. The success or failure of the commission now hung precipitously on the shoulders of a single overworked junior staff lawyer.

The second area, the identity of the assassin, was assigned to Joseph Ball and David Belin, two experienced trial lawyers. It was their job to assemble and weigh the evidence that established the identity of the assassin. However, the identity of at least one assassin had already been predetermined since Albert Jenner and Wesley Liebeler were assigned to the third area of investigation, namely, Oswald's background and the events in his life that caused him to assassinate the president.

This investigation was further enhanced by William Coleman, a former special counsel for the city of Philadelphia, and W. David Slawson, a Denver lawyer, two men experienced in working with federal agencies who were assigned to the fourth area, Oswald's "possible conspiratorial relationships." They were particularly interested in Oswald's movements abroad and whether he could have been influenced to act by a foreign power.

The fifth area of inquiry, Oswald's death, was assigned to two former United States attorneys, Leon Hubert and Burt Griffen. Their primary areas of concern were principally whether Jack Ruby might have had help or inducement to murder Oswald, and whether Ruby knew Oswald prior to the assassination.

A sixth area of inquiry was later added at the request of the commission. It consisted of investigating the political hot potato of the president's protection in Dallas. This task was assigned to Samuel Stern, an attorney from Philadelphia, with Lee Rankin acting as the senior counsel in this area. These men would take a long look at the precautions taken by the Secret Service and the FBI to protect Kennedy's motorcade.

THE WARREN COMMISSION

On January 13, 1964, the FBI supplemented its original 384-page report with a 67-page document that addressed Oswald's history, associates, and affiliations. They also forwarded to the Warren Commission a 39-page report addressing the omissions in the first report concerning Jack Ruby, together with a 26-page supplement discussing Ruby's murder of Oswald. These reports concluded that Ruby and Oswald had no prior association and that Ruby acted alone when he shot Oswald. Both documents filled in some of the vital details that Warren Commission members felt were missing from the first FBI report.

However, by January 22, 1964, two months after the assassination, a serious conflict of interest arose involving the FBI. Vincent Wagonner Carr, the attorney general of Texas, informed the commission's general counsel, Lee Rankin, by phone that an allegation had arisen in Texas that Lee Harvey Oswald had been an "undercover agent," or a "paid informant," for the FBI. This allegation threatened to undermine the investigative work of the FBI concerning the assassination and invalidate the agencies' conclusions concerning Oswald.[10]

Chief Justice Earl Warren called an emergency session of the commission. When Rankin laid out the Texas allegations before the members, they immediately recognized the potential repercussions of this bolt from the blue. A linkage between Oswald and the FBI could be interpreted as a conspiracy with a government agency to assassinate the president. In such a case, the FBI would have an overriding interest in assuring the commission and the public that Oswald acted alone. Even though the allegations were little more than speculations, if the rumor reached the public, the damage could be irreparable. This single accusation would plague the commission for months and drain off valuable time from its work. By January 27, the commission concluded that it couldn't just accept the FBI's word that Oswald was not a paid informant for them and therefore decided to conduct its own investigation into the matter.

Meanwhile, in the midst of this sideshow, the commission had to set about the work of determining the truth of what happened at Dealey Plaza. But most of the commission's lawyers agreed that the investigation could not start in earnest until after Jack Ruby's murder trial was concluded. Ruby's legal rights could have been compromised if the commission's investigations were to commence prior to the conclusion of his murder trial. Therefore, it wasn't until March 14, when the Ruby trial ended with a guilty verdict, that the investigation began. This put the investigators under very tight time pressure because Wesley Liebeler had requested reports by June 1, 1964.

On March 16, Arlen Specter traveled to Dallas to conduct field investigations. His primary concern initially was to address the issue of the president's throat wound. Earl Warren considered this issue crucial because of the rumors surrounding its origin. The Dallas doctors had indicated that the throat wound was a wound of entrance. This would place a shooter in front of the president, contrary to initial views that all shots came from Oswald's rifle from the Texas School Book Depository, positioned behind the motorcade at the time of the shots.

Over the course of the following week, Specter interviewed twenty-eight Dallas doctors and medical personnel. All of Dallas's Parkland Hospital doctors who saw the throat wound conceded to Specter that the wound could have been either an entrance wound or an exit wound. He discovered that the original rumor started with an answer Dr. Malcolm Perry, one of Kennedy's attending physicians, had given to a question. Dr. Perry told Specter that he had no basis to conclude that the wound was an entrance wound and told the commission that he only said that the throat wound *could* have been a wound of entrance.[11]

However, the pernicious problem of multiple shooters in Dealey Plaza had already raised its ugly head before the commission in another manner. Frame-by-frame analysis of the Abraham Zapruder film, taken by a Dallas businessman with his home-movie camera at the precise moment of the assassination and made available to the commission by January 27, 1964, showed that the latest point that Kennedy could have been hit was in film frame 225 (Z225), because Kennedy is clearly seen to be wounded as he emerges from behind the visual obstruction of a road sign while his motorcade moves along Elm Street. A Secret Service reenactment of the assassination showed that Oswald's view from the Texas School Book Depository would have been blocked by a large oak tree between frames Z166 and Z207.[12] Based on eyewitness testimony regarding the timing of the shots, it became clear that the first must have been fired sometime after Z207, when the assassin had a clear view of Kennedy. However, by February 25, 1964, an original high-quality Zapruder print of the film was made available to the commission. It clearly showed that Governor Connally was in distress, and therefore definitely hit, by frame Z243, but he likely was struck sooner than this as his doctors asserted that he was not in a position to be hit after frame Z240 since his wound was in his back and he was turning sideways toward the president during these frames.[13]

A serious problem had now arisen. Live fire tests conducted by the FBI on the Mannlicher-Carcano rifle found in the Texas School Book Depository area where the fatal shots were alleged to have been fired showed that the bolt could not be operated any faster than 2.3 seconds per round. However, the Zapruder film ran at 18.3 frames per second. At absolute maximum only 36 frames, or 1.97 seconds, had elapsed between the firing of the shot that first struck Kennedy and the bullet that struck Connally. Therefore, two shots could not have been fired by Oswald in the time frame required to hit both Kennedy and Connally with separate bullets.

There was only one way out of this dilemma. The reactions of both Kennedy and Connally exhibited on the Zapruder film could potentially be explained if both men had been struck by the same bullet. This hypothesis would become known as the single-bullet theory and was first advanced to the commission by Arlen Specter on March 16, 1964.

Specter had carefully interviewed the autopsy doctors before advancing his hypothesis. Commander James Humes, the lead doctor at the Kennedy autopsy, had volunteered that although he could not find a path for the bullet through Kennedy's body, he had concluded deductively that the bullet must have passed through his body and exited the throat. Humes told the commission, "I see that Governor Connally is sitting directly in front of the late President, and suggest the possibility that this missile, having traversed the low neck of the late President, in fact traversed the chest of Governor Connally."[14]

Figure 1: Z243. Kennedy and Connally in distress. Kennedy can be seen clutching his throat in response to being struck there by a bullet while Connally's face appears contorted in pain. Zapruder film © 1967 (renewed 1995). The Sixth Floor Museum at Dealey Plaza.

On April 27, 1964, Dr. Alfred Olivier, a veterinarian and army ballistics expert, conducted live-fire tests on the alleged murder rifle, which were supervised by Arlen Specter. Because Specter thought it would be too difficult and complex to fire through multiple targets simultaneously, the tests were conducted on single targets only. Shots were fired through gelatin blocks, through an anesthetized goat, and through the wrist of a cadaver. The rounds fired directly into the cadaver's wrist produced a substantial amount of damage, greater than that which had been inflicted on Connally's wrist. When a bullet passes through an object, it loses energy and velocity due to frictional forces. As it slows down, it becomes less able to generate damage when it impacts a target. As a result, Dr. Olivier concluded, "My feeling is that it would be more probable that it [the bullet that struck Connally's wrist] passed through the president first."[15]

A key element of the single-bullet theory involved the bullet itself. CE (Commission Exhibit) 399 was an intact bullet discovered on a stretcher in Parkland Hospital. Specter believed that this bullet had been found on Connally's stretcher and therefore was the bullet that had passed through both men.[16]

The commission would place a high standard on the requirements for reliability of eyewitnesses. Although the commission heard testimony from twenty witnesses who claimed to have heard shots being fired from the grassy knoll, their testimony was discounted because it was unsupported by the physical evidence. Wesley Liebeler dismissed eyewitness testimony out of hand as unreliable unless it was supported by substantial physical evidence.[17] No rifles or shell casings were found behind the fence on the grassy knoll. Further, the possibility of a shooter at this position was precluded by the medical evidence from the official autopsy, which clearly showed that the shots were fired from a position behind Kennedy. In such a case the commission attorneys reasoned that the autopsy evidence of the wounds on the body constituted the "best evidence," and so it was believed to supersede eyewitness testimony of shots coming from in front of the motorcade.

The writing of the commission's report began in earnest after the June 1, 1964, deadline for its completion ended. By this time, most of the teams had not even completed their field investigations. To accelerate the writing of the report, the commission's general counsel, Lee Rankin, appointed Norman Redlich, Alfred Goldberg, and Howard Willens as a reediting committee.[18] Although Arlen Specter had submitted his report by the deadline, it had to be rewritten by Redlich because the commission's members held lingering doubts about the single-bullet theory.

The first chapter, written by Redlich, was a brief summary of the event. The second chapter was a more-detailed explanation and outline of the assassination, again written by Redlich with some help from Specter and Samuel Stern. The third chapter dealt with the basic facts, the source of the shots, the timing, and the medical evidence. This was Specter's area. The fourth chapter focused on Lee Harvey Oswald and the evidence that tied him to the crime. It was originally written by Joseph Ball and David Belin, but as in the case of Specter's chapter, the content was substantially rewritten by Redlich.[19] Chapter four effectively presented the case against Oswald.

The fifth chapter concerned itself with Oswald's death at the hands of Ruby. Since the original members assigned to this task, attorneys Burt Griffen and Leon Hubert Jr., were still knee-deep in their field investigations into July, other lawyers were assigned to write their chapter. Chapter 5 was originally drafted by Burt Griffen and Murray Laulicht, but was completely rewritten by Alfred Goldberg.[20]

The sixth chapter also focused on Oswald. Its goal was to examine Oswald's movements overseas. It was originally drafted by Stuart Pollack and David Slawson, but was also ultimately rewritten by Goldberg in an attempt to deal with the issue of alleged conspiracies and rumors.

The seventh chapter dealt with Oswald's associates and motives. Wesley Liebeler originally drafted this chapter, but it was also essentially rewritten by Goldberg. The last chapter addressed the problem of Kennedy's protective services. This chapter was drafted by Samuel Stern but later substantially rewritten by Howard Willens.[21]

The reports were forwarded to the commission members for review only after they had been reworked by the reediting committee. The commissioners provided extensive comments and returned the drafts to the committee for rewriting. Some chapters were rewritten as many as twenty times.[22] The commissioners' input into the final report was therefore reflected in this process.

The report itself had to be supplemented by twenty-six volumes of exhibits and testimony. The final draft of the report was completed by the middle of September 1964.

THE WARREN COMMISSION

The Warren Commission report was released on September 28, 1964. It reached the following conclusions:

(1) The shots that killed President John F. Kennedy and wounded Governor Connally were fired from the sixth-floor window at the southeast corner of the Texas School Book Depository (TSBD).

(2) The weight of the evidence indicates that there were three shots fired.

(3) Although it is not necessary to any essential findings of the commission to determine just which shot hit Governor Connally, there is very persuasive evidence from the experts to indicate that the same bullet that pierced the president's throat also caused Governor Connally's wounds. However, Governor Connally's testimony and certain other factors have given rise to some difference of opinion as to this probability, but there is no question in the mind of any member of the commission that all the shots that caused the president's and Governor Connally's wounds were fired from the sixth-floor window of the Texas School Book Depository.

(4) The shots that killed President Kennedy and wounded Governor Connally were fired by Lee Harvey Oswald.

(5) Oswald killed Dallas Police Patrolman J. D. Tippit approximately forty-five minutes after the assassination.

(6) Within eighty minutes of the assassination and thirty-five minutes of the Tippit killing Oswald resisted arrest at the theater by attempting to shoot another Dallas police officer.

(7) The commission has found no evidence that either Lee Harvey Oswald or Jack Ruby was part of any conspiracy, domestic or foreign, to assassinate President Kennedy.

(8) In its entire investigation the commission has found no evidence of conspiracy, subversion, or disloyalty to the US government by any federal, state, or local official.

(9) On the basis of the evidence before the commission, it concludes that Oswald acted alone.

HEAD SHOT

But even before the commission's final report was made public, one if its own members privately doubted its conclusions.

> LBJ had tried to telephone him [Senator Richard Russell] all afternoon [September 18, 1964]. Russell, at home in Georgia, finally returned the call. The conversation was remarkable. Neither Russell nor Johnson believed a single bullet could account for all seven of the nonfatal wounds of President Kennedy and Governor Connally—the crucial conclusion at the heart of the commissions' Report: "Well, I don't believe it," said Russell.
> "I don't either," Johnson replied.[23]

CHAPTER 2

EDWARD EPSTEIN
THE SINGLE-BULLET THEORY CHALLENGED

The critical problem in government investigations is inherent in the evaluation, not the accumulation, of data. The fact that the government can amass a virtually unlimited amount of information on any given subject only intensifies the problem.

—Edward Jay Epstein, *Inquest: The Warren Commission*
and the Establishment of Truth

Edward Jay Epstein was the first person to criticize the conclusions of the Warren Commission report in print. His book *Inquest: The Warren Commission and the Establishment of Truth* took a critical look at the inner workings of the commission, how it performed its investigations, and how it went about reaching its final determinations. Epstein also decisively examined the commission's conclusions in light of the data published in the report, supplemented by personal interviews. Epstein's book was based on his master's thesis in political science at Cornell University and as such commanded credibility due to its academic origins and professorial scrutiny.

Epstein believed that the commission was compromised from the outset because of the fundamental problem posed by its dual purpose. The initial executive order establishing the commission never adequately set forth its purpose. It listed the goals as examining evidence, conducting investigations, and evaluating facts but never stated why this was to be done. Chief Justice Earl Warren stated, "The purpose of this Commission is, of course, eventually to make known to the president, and to the American public everything that has transpired before the commission."[1] This statement reduced the commission to a simple fact-finding body without motivation. However, Warren had accepted the appointment because Johnson had made it clear to him that "the nation's prestige was at stake." John McCloy said that it was of the utmost importance to "show the world that America is not a banana republic, where a government can be changed by a conspiracy."[2] Congressman Gerald Ford told Epstein that dispelling damaging rumors was a major concern of the commission, a statement with which most of the other members agreed.[3] Epstein concluded:

> There was thus a duality of purpose. If the explicit purpose of the commission was to ascertain and expose the facts, the implicit purpose was to protect the national interest by dispelling rumors. These two purposes were compatible so long as the damaging rumors were untrue. In a conflict…one of the commission's purposes would emerge as dominant.[4]

Although fact finding was the stated goal, an implicit undercurrent tainted the investigation. The motivation for the committee members was something other than strictly determining the truth. Their motivation was rehabilitation of the nation's reputation in light of a possible coup d'etat. A nation's reputation among other nations is key to its effectiveness in the international arena, both militarily and economically. While it may not be entirely fair to criticize the commission from the enlightened vantage point of the twenty-first century, at the same time it would be foolish to imagine that men whose goal was something other than the attainment of fact would seek truth unerringly. In any review of history, it is always necessary to know the viewpoints, the motivations, and the aspirations of the historian who recorded it. Thus, Julius Caesar's *Commentaries on the Gallic Wars* can only be evaluated in light of his desire to look good before the Roman people. His work is a primer on the historical art of propaganda. The key is to tell the audience only information that supports the goal of the work, and to selectively withhold facts that do not.

Part of this problem stemmed from the way the commission had chosen to go about its work. It had decided early on to rely on the investigations of federal agencies without hiring independent bodies to bring facts to light. The commission thought that its own lawyers were all the independent investigators that were required to do the job. In theory, this approach had the potential to be successful, but in practice critical problems arose stemming from the overwork and overuse of their limited participating legal staff and from interactive issues with the agencies they tasked to do their work.

As an example, crucial evidence of a second possible assassin uncovered by the FBI never reached the commission. The FBI had interviewed a Mrs. Eric Walther in early December of 1963. She recalled that she was standing across the street from the Texas School Book Depository when she saw a man in the window with a rifle, whom she described as having blond hair or light hair and wearing a white shirt. In the FBI's report of the interview, her description was recounted:

> This man was standing in or about the middle of the window. In the same window was another man.... The other man was standing erect and his head was above the open portion of the window. As the window was very dirty she could not see the head of the second man. ... This second man was apparently wearing a brown suit coat, and the only thing she could see was the right side of the man, from about the waist to the shoulders. Almost immediately after noticing this man with the rifle and the other man standing beside him, someone in the crowd said "Here they come..."[5]

Further, this account was corroborated by a second witness, Arnold Rowland. Rowland testified that he had also seen a rifleman in the sixth-floor window of the Texas School Book Depository and a second person on the same floor.

> I noticed on the sixth floor of the building that there was a man back from the window, not hanging out the window. He was standing and holding a rifle. This appeared to me to be a fairly high-powered rifle

because of the scope and the relative proportion of the scope to the rifle, you can tell about what type of rifle it is.[6]

However, although the FBI report narrating Mrs. Walther's testimony was submitted to the commission and was included in commission document #7, it was never actually evaluated by the commission. She was never called to testify and was never interrogated by staff lawyers. Rowland's corroborative testimony was rejected by the commission and not used in the report. Although the FBI was submitting material that potentially indicated the involvement of multiple people in the assassination, this information was somehow not reaching the commission's collective consciousness.

A further problem stemmed from how the FBI operated in its role as investigative arm of the commission. The FBI was very good at finding information at the commission's behest, but agents limited the scope of its inquiry strictly to the facts and information actually requested. For instance, Rowland's account never appeared in any of the FBI's reports. Epstein believed that the reason for this was that the FBI simply focused exclusively on the specific issue it was investigating to the exclusion of all else. In this case, the FBI was investigating the specific issue of whether Rowland could identify the man he saw in the window carrying a rifle. A second man on the floor with the gunman was not an object of their inquiry and consequently was ignored.

Epstein's most substantive criticisms of the commission, however, centered on the controversial single-bullet theory. This theory was originated by Specter as a way out of the dilemma posed by the timing of the gunshots shown in the Zapruder film and corroborated by witnesses.

Careful analysis of the Zapruder film showed that the assassination could have been committed by one man only under one condition: that Kennedy and Connally were hit by the same bullet. The spacing of the two shots as determined by reaction correlations on the film was less than the 2.3 seconds required to operate the bolt on the suspected murder weapon. However, both the FBI summary and its supplementary reports relating to the autopsy stated that the first bullet did not exit from the front of the president's body. "Medical examination of the president's body had revealed that the bullet entered his back and penetrated to a distance of less than a finger length."[7] Therefore, no path through Kennedy's body was ever determined for the bullet that inflicted this back wound at the official autopsy. The information that was determined at the autopsy was that the wound appeared to be at a downward angle of between 45 and 60 degrees. The reports of the autopsy were therefore very inconsistent with the possibility of a bullet having entered Kennedy's back and exiting from his neck.

However, these reports of the autopsy were found to be inconsistent with the official autopsy report itself. Commander James Humes, who led the official autopsy, testified before the commission and submitted an affidavit to the effect that he subsequently burned his "preliminary notes" taken at the autopsy. His conclusions in the official report radically conflicted with the descriptions in the FBI summary and supplementary reports. In his official report of the autopsy, Humes states that the bullet hit the president in the rear of the neck and exited through the throat.

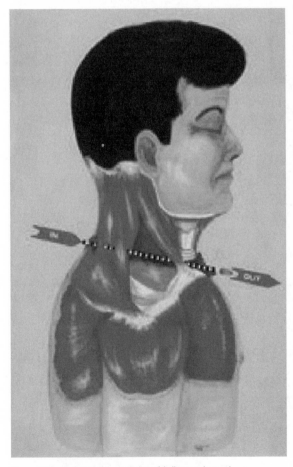

Figure 2: Warren Commission conclusion of bullet entering neck.

Even if the FBI reports were somehow inaccurate, the conclusion that the president was hit in the rear of the neck was challenged by other evidence that called the entire single-bullet theory into question. The photograph of the president's shirt shows that the bullet entered his body below the collar line, which is consistent with the FBI summary report's description of a wound "just below his shoulder," but inconsistent with the commission's description of a wound in "the rear of the neck." Epstein noted that a bullet entering the president's back six inches below the collar line could not have conceivably exited through his throat. The autopsy surgeons' inability to find a path for the bullet is further evidence

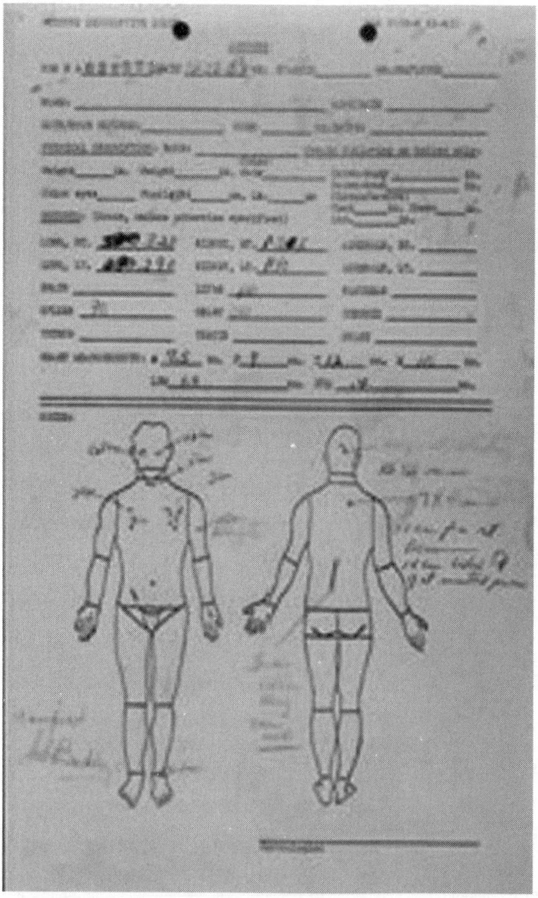

Figure 3: Autopsy sheet prepared by Dr. Boswell showing the wound in Kennedy's back, CE 397.

that the bullet simply did not pass completely through the president's body. Secret Service agent Roy Kellerman, who was present at the autopsy, reported, "Colonel Finck—during the examination of the president...he is probing inside the shoulder with his instrument.... He said, 'There are no lanes for an outlet of this entry in this man's shoulder.'"[8]

Dr. Boswell's autopsy sheet also shows the wound appearing in Kennedy's back, not the rear of his neck (figure 3). Dr. Boswell assisted Dr. Humes at the official autopsy in Bethesda, Maryland. This sheet contains original bloodstains.

A further problem with the single-bullet theory stems from the testimony of Secret Service agent Kellerman, who was sitting in the front seat of the president's limousine. Kellerman testified that he distinctly heard the president say, "My God, I am hit," after the first shot. If the first shot had hit Kennedy in the throat, puncturing his windpipe, it is medically highly implausible that he could have spoken after receiving this wound. This is completely inconsistent with the commission's version of events suggesting that the *first* bullet to strike Kennedy exited through his throat.[9]

Other testimony corroborated this account. Both Connally and his wife believed that Kennedy and Connally were struck by separate bullets. Connally testified that he actually heard the first shot that was fired but could not hear the shot that he felt strike him.[10] A Mannlicher-Carcano bullet travels at about 2,000 feet per second, about twice the speed of sound. Connally believed that the shock of being struck by the bullet prevented him from hearing the sound of the shot that would have followed it. Since he had actually heard the first shot, the one that had struck Kennedy in the back, Connally was convinced that he must have been hit by a second shot.

Connally's wife, Nellie, agreed with him. She testified before the commission that "the president had both hands at his neck," after she heard the first shot.[11] Then, a few seconds later, she heard the second shot that hit her husband. Seated next to her husband and directly to the front left of Kennedy, Mrs. Connally had a superior firsthand observation of both men, their reactions, and the sounds of rifle shots as they were fired. No one was in a better position to view the sequence of responses of the two men being wounded by rifle fire, coupled with the aural timing of the shots. Mrs. Connally was convinced, as was her husband, that the two men had been wounded by separate bullets.

The single-bullet theory had another problem. Commission Exhibit (CE) 399, the pristine bullet, was found on a stretcher at Parkland Hospital. For the single-bullet theory to hold up, this spent round must have been found on Governor Connally's stretcher. This was the position argued by Arlen Specter, that this round caused both the back and the throat wounds to Kennedy and the back and front wounds to Connally, together with the injuries to his wrist.

But Specter's conclusions hinged on Oswald firing just three shots. One shot, he concluded, hit both Kennedy and Connally. One shot hit Kennedy in the head, and a third shot hit the curb, ricocheted, and struck a bystander, James Tague.[12] Consistent with this conclusion was that precisely three spent shell casings were found in the sniper's nest of the sixth-floor window of the Texas School Book Depository. Bullet CE 399 was ballistically matched to the rifle found in the sixth floor of the depository. If CE 399 had been found on Connally's stretcher, then it could have been the bullet that struck both men. But if it were found somewhere else, it posed a problem.

When Arlen Specter actually interviewed the witnesses at Parkland Hospital, days after he first advanced his conjecture to the committee, he found evidence that conflicted with his own theory. The man who had actually found the bullet, Darrell Tomlinson, a hospital engineer, told him that the bullet had specifically not come from Connally's stretcher, but instead had come from the stretcher parked in front of him in the emergency room.[13] It was entirely possible that this stretcher could have been Kennedy's. If CE 399 was not the bullet that had wounded Connally, then a fourth shot must have been fired from somewhere.

Further, Army Lt. Col. Pierre Finck, an expert in wound ballistics, testified that he thought that CE 399 could not have been the bullet that caused Connally's wrist wound because more bullet fragments were removed from the wound than were lost from the bullet itself,[14] a conclusion supported by the other medical experts who testified before the commission. Damaged bullet fragments found inside the limousine were too distorted to be ballistically matched to Oswald's rifle; however, these fragments were believed by the commission to have been from the shot that struck Kennedy in the head. If CE 399 did not inflict the wounds on Connally, then another round must have. This means a fourth shot. Based on eyewitness testimony and the number of casings found in the sniper's nest, the commission concluded that only three shots were fired by Oswald. Therefore, rather than support the single-bullet theory, the pristine bullet presented a problem for the entire scenario involving Oswald as the sole assassin.

Other serious problems arose when actual live-fire tests were conducted with Oswald's Mannlicher-Carcano rifle. The first tests with this weapon were conducted by the FBI as early as November 27, 1963.[15] Three marksmen fired three shots each at a target fifteen yards away. None of the experts could fire off the three shots in less than 6 seconds, with one taking as long as 9 seconds. Analysis of the Zapruder film showed that 5.6 seconds had elapsed between the earliest possible aimed shot and the final shot that struck Kennedy in the head. Worse still, all the shots were high and to the right of the aiming point. Oswald's target had been a hundred yards away; theirs was at a distance of only fifteen yards. Oswald would have had to have been a spectacular marksman to have accomplished the assassination with this weapon in the 5.6-second window at such a range.

A second series of tests conducted by one of the marksmen, Robert Frazier, fared no better. He still could not reach the 5.6-second mark for firing three shots, and all of the shots missed the target, five inches high and five inches to the right. Frazier would later testify before the commission that inaccuracy of the weapon was due to an uncorrectable deficiency in the telescopic sights.[16]

A third series of tests was conducted by the Army Ballistic Research Laboratory in Aberdeen Proving Ground, Maryland, on March 27, 1964. On this occasion, three certified expert marksmen fired shots at silhouette targets over the same range that Oswald fired. These experts were able to exceed the 5.6-second time barrier with three-shot sequences as low as 4.6 seconds, but all missed targets, sometimes substantially, in the process.[17] However, this test did not take into account aiming time. Oswald would have had to have fired the instant that Kennedy's limousine had emerged from the cover of the oak tree. The marksmen had unlimited time to aim their first shot. Second, the experts were not firing at moving targets. Oswald was.

The salient problem with this test, however, was that to accomplish it the army experts had found it necessary to add "three shims," small metal plates, to the telescopic sight to correct its accuracy. Otherwise, the weapon could not be aimed accurately. This raises the question of how Oswald could have surpassed the marksmanship of such experts, who were firing in a staged test, using a telescopic sight that was defective and uncorrected. If the marksmen could not use this sight uncorrected to hit targets over the range from the Texas School Book Depository to Kennedy's limousine, how could Oswald have done it?

Not surprisingly, perhaps, the original version of the chapter 4 report submitted to Wesley Liebeler contained no mention of the problems with the scope. Redlich had essentially rewritten this chapter many times in an attempt to satisfy the commissioners. However, in a caustic twenty-six-page memorandum, Leibeler detailed a series of criticisms with this chapter that particularly pointed out the problems of "omission." He felt that leaving out a discussion of the difficulties with the telescopic sight was "simply dishonest." However, the final correction to this problem was that while the trouble with the sight was inserted into the chapter, it was insinuated that the defective sight actually enhanced Oswald's accuracy for a moving target.[18]

Because of the problems with the single-bullet theory, and the difficulties in live-fire tests with the actual rifle found in the depository, Senator Richard Russell of Georgia became suspicious of the theory of Oswald as the sole assassin. He consequently consulted a former army intelligence officer, Colonel Phillip J. Corso.

Phillip J. Corso was a controversial figure. During the Eisenhower administration, he served on the National Security Council. In 1961, he became chief of the Pentagon's foreign technology desk in army research and development, under Lt. Gen. Arthur Trudeau. Two years later, he left the army and became an aide to Senator Strom Thurmond. However, in 1997, he wrote a book on UFOs titled *The Day after Roswell* in which he claimed to have seen alien bodies and in which he purported to have run a program to reverse-engineer alien technology discovered at Roswell, New Mexico.[19] This account was not well received by the news media and was the subject of considerable criticism and controversy. The *Guardian* listed it among the year's top-ten literary hoaxes.[20] Though this work is surprisingly lucid and the discussion logical and well structured, it is important to realize that Corso was an intelligence officer. It must always be kept in mind that one of the roles of intelligence is to "disinform." Corso's controversial contentions may have been made in order to mask the true nature of classified programs with which he was associated.

However, Colonel Corso claims to have told Russell in 1964 that the Mannlicher-Carcano rifle did not have the necessary range and was far too inaccurate to have been Kennedy's murder weapon. Corso further explained to Russell that it was likely that an accurate small-caliber high-velocity weapon was employed to inflict the wounds on Kennedy.[21] Based on these conversations, Russell became even more suspicious of the conclusions that were being advanced and advocated by the junior investigators for the commission.

Epstein reported that not only did Senator Russell disagree with the single-bullet theory, but also that Senator John Cooper of Kentucky and Representative Hale Boggs of Louisiana disagreed with it as well.[22] Cooper said, "I too objected to such a conclusion; there was no evidence to show that both men were hit by the same bullet." Boggs said, "I had strong doubts about it [the single-bullet theory]," and added that he felt the question was never resolved.

The Warren Commission failed in its goal of elucidating the truth of the assassination. It couldn't even find a conclusion with which its own members could agree. This, perhaps, is the very definition of failure.

The essential problem of the Warren Commission, the liability that doomed it even before it started its work, was that it was a homogenous organization. Every member was a lawyer. When a group lacks diversity of backgrounds, the variety of the input from its members and the range of that input are limited. This was the crucial drawback that doomed the commission right out of the gate.

Contrasted against this approach is that used by the Rogers Commission, which was established in 1986 to investigate the *Challenger* Space Shuttle explosion. The commission chose Professor Richard Feynman of Caltech to serve on the committee to investigate the tragedy. Many believed that his selection was as convincing as anything else that the committee was intent on getting to the truth of the disaster. In addition to enjoying an enormous reputation in the scientific community, Feynman was known as a straight-shooter who was moved by nothing except a quest for truth. This pursuit had been the hallmark of his career and the key to his prodigious success as a scientist. As a member of the Rogers Commission, which was headed by former attorney general Richard Rogers and included current and former air force generals and astronauts like Neil Armstrong and Sally Ride, Feynman actually conducted his own independent investigation and prepared his own report, which was included as an appendix to the official report of the committee.

Scientists and engineers, like Feynman, understood that if they didn't conduct a thorough investigation, if they didn't find out what really happened, if they didn't get to the bottom of the problem, then the space shuttle would be in danger of blowing up again the next time it was launched. The next time it took flight, the problems wouldn't be corrected and the shuttle program would have the same result, a billion-dollar spacecraft in pieces, tragic loss of life, and a program in shambles. Insanity is doing the same thing over and over again and expecting a different result.

The same argument can be applied to the Warren Commission. If the commission didn't get to the bottom of the assassination, if they didn't find out what really happened and what triggered the tragedy, then something like it could and probably would happen again. The consequences of failure for the commission were large indeed. Not only would the public lose confidence in government, but another president could once more be cut down in his or her prime.

Most scientists and engineers swear by Murphy's law and all its corollaries. One of the corollaries of Murphy's law (*If it can go wrong, it will go wrong*) is that if a sequence of events has the potential to do more damage than alternative possible sequences, the sequence that does the greatest possible harm will be the one that actually occurs. Certainly, this is what happened to the Space Shuttle *Challenger*. Another corollary is that if there is a chain of damaging events that you didn't anticipate and a sequence that you did anticipate, the sequence of events that happens will be the one you didn't anticipate. My favorite corollary of Murphy's law is *Murphy was an optimist.*

Ultimately, unlike the diverse and successful Rogers Commission, the homogenous makeup of the Warren Commission doomed it from the start. There was a crucial stake at every level and success at finding the truth could only be achieved through a diversity of opinions.

This was something Kennedy himself realized as a result of the Bay of Pigs disaster. No one had disagreed when the CIA presented its plan to invade Cuba with a band of expatriates. No one dissented. Everyone was in accord. This was not a recipe for success. In order to make meaningful decisions, it was essential to be able to evaluate alternatives. Alternate choices can only arise from a group of people who have a variety of viewpoints and a multiplicity of backgrounds, worldviews, and fundamental beliefs. Kennedy took steps to surround himself with such people as insurance against a disaster like the Bay of Pigs ever plaguing his administration again. It is notable that this strategy may have saved his presidency and the nation when the Cuban Missile Crisis arose. It was Kennedy's deft handling of this delicate predicament that defined his presidency.

If the commission wanted to get at the truth of Kennedy's death, it needed people with varied backgrounds in life to try to provide a plurality of possibilities and scenarios. Instead, it had the opposite.

HEAD SHOT

When I was a contractor for the navy, working for the Johns Hopkins Applied Physics Laboratory in Laurel, Maryland, I participated in a test fire for a submarine-launched missile. The navy captain who was in charge of the launch lined everyone up in a single conference room, the navy people on one side and the technical contract specialists on another. We then debated which missile to launch. This was the way to get at the truth. The discussion went back and forth, and the debate at times got heated, but the captain could listen to both sides, no matter how polarized, and through this rigorous vetting process arrive at the right answer. Everyone was given a fair opportunity to express viewpoints and rebuttals to the deciding authority.

When I worked at the Naval Research Laboratory (NRL) in Washington, DC, I wanted to publish a paper in a controversial scientific area. The associate director of the NRL was concerned about the laboratory's reputation and as a result I was asked to defend the paper before a panel of experts composed of scientists throughout the lab. This sort of defense was above and beyond the normal reviewing procedure for publications, but it was deemed necessary because of the potential attention such a paper might receive. Because the work touched on a number of fields in diverse scientific disciplines, scientists from all over the Condensed Matter and Radiation Sciences Division were asked to review the paper prior to submission. I was required to defend my paper to a room full of scientists who specialized in solid state physics, nuclear physics, materials science, and other domains. Everyone had an opportunity to express his viewpoint of the work and its implications and to receive an immediate response to the comments. Even though the associate director could not possibly have had a background in all those fields, he did have enough of a technical background to follow the discussion. As a result of my successful defense of my work before this panel, in full view of the associate director, I was allowed to submit my paper for publication.

This was precisely the type of protocol Kennedy followed in his own administration. He started putting people on his cabinet who had different and opposing viewpoints and disparate backgrounds so that he would have dissenters and multiple points of view. This allowed him to listen to opposing arguments and have ideas vetted before implementing them. Kennedy was determined never to have a disaster like that in Cuba plague his presidency again.

However, nothing like this ever occurred on the Warren Commission. They lacked a fully interactive vetting process, and even if they had attempted to implement one they lacked the technical expertise to effectively conduct it considering the complex ballistic and medical evidence before them. But any vetting process would have been better than none at all. Instead, attorneys wrote their reports, submitted them, and then responded to comments. No one ever called them all together and tried to listen to opposing viewpoints in an interactive manner in an effort to reach the truth of the assassination. This was a major failing of the commission and was due in no small part to its homogenous makeup.

This homogeneity and lack of a gadfly hampered the report's final conclusions. A skeptical, exhaustive thinker like Professor Feynman, had he been on the commission, would have prevented the other members from concluding embarrassing absurdities like Oswald's misaligned scope actually helping him to assassinate the president. It was a difficult enough of a feat for a marksman to hit a moving target at 270 feet three times in a six-second period. If he has to adjust for an offset in open space, this would make the exploit far more difficult. If the assassin has to do anything other than line up the crosshairs on the target, both his time and his accuracy would be compromised. A badly misaligned scope could only have been a detriment to Kennedy's killer, and could in no way have facilitated his ability to accomplish his gruesome task.

Professor Feynman, or someone like him, would have prevented the Warren Commission from publishing contradictory conclusions like "There is very persuasive evidence from the experts to indicate that the same bullet which pierced the president's throat also caused Governor Connally's wounds. However, Governor Connally's testimony and certain other factors have given rise to some difference of opinion as to this probability but there is no question in the mind of any member of the commission that all the shots which caused the president's and Governor

Connally's wounds were fired from the sixth floor window of the Texas School Book Depository," coupled with "The shots which killed President Kennedy and wounded Governor Connally were fired by Lee Harvey Oswald." If all the commission members didn't believe the single-bullet theory, then they didn't believe that all the shots were fired by the same person. No one man could accomplish this feat because the bolt of the purported murder weapon could not be operated fast enough. Unless the single-bullet theory was valid, there had to be multiple shooters in Dealey Plaza. Doubts in one key aspect of the facts of the assassination placed doubts about the entire final determination of the events in Dealey Plaza. Therefore, these conclusions were self-contradictory.

At minimum, a final report must be internally consistent. The conclusions can't contradict themselves if the report is to remain credible. Saying that not all commission members agree with the single-bullet theory but all agree that only one shooter inflicted the wounds is akin to publishing the Epimenides Paradox as a conclusion:

> *The statement below is true.*
> *The statement above is false.*

This is simple logical absurdity. This failing is so serious that it invalidates the conclusions of the commission. They couldn't reach a consensus and, worst of all, they refused to admit it, thereby relying on illogicality instead. The commission should have continued its work until a valid working consensus was reached. It didn't. Logical conundrums and absurdities did not help the report's credibility with the public.

The second major problem was the creation of a bottleneck. All activity of the investigators and all reports were first sent through Lee Rankin before reaching the committee.[23] So if this single individual did not place the arguments properly or if they were too filtered in one direction, the truth would be compromised. This effectively negated the advantage of having a committee because everything was going through just one individual to reach that committee, filtered and distorted through the lens of a single man's perspective. No one was willing to sidestep Rankin and do his own independent investigation as Feynman did for the Rogers Commission.

This problem was exacerbated by assigning the task of ascertaining the basic facts of the assassination, the key to the entire investigation, to one man, Arlen Specter. This assignment would have been impossible for ten men to perform adequately. This forced Specter to focus on limited aspects of the assassination, like investigation of Kennedy's throat wound. Specter spent eight days in Dallas interviewing medical personnel, and except for his discussions with Jean Hill, a grassy knoll witness, and Jack Ruby, this constituted the entirety of Specter's field investigations of "the basic facts of the Assassination."[24]

How can one person hope to arrive at the right answer when he has spent almost all his investigative time on a single issue? The commission essentially had one person doing all the work. Specter couldn't possibly do all the research himself. If the key person is overworked, he can't reach the right conclusions on the series of questions he is supposed to investigate. The other problem is that only one viewpoint and one paradigm of analysis is operating when many different viewpoints and alternative paradigms are needed so that a consensus among many people and perspectives is achieved and not just the synthesis of one man. It is a dangerous liability for any committee wishing to find the truth to leave the heart of the investigation essentially in the hands of one person, because this person's opinion will carry weight over everyone else since he's done the hands-on research. He's most familiar, or perceived to be most familiar, with the actual facts. This strategy is fine so long as this person reaches correct conclusions, but if he finds erroneous ones, then it's a disaster. Since Specter could not possibly have completed all the necessary research in this area himself, it's highly unlikely he could have reached a correct solution based on the available data. Assigning a single person to an overwhelming area of investigation was therefore a recipe for failure and of ultimately reaching the wrong conclusions.

As if this were not enough funneling, the final report itself ended up being primarily the work of just two men, Norman Redlich and Alfred

Goldberg.[25] Although as many as thirty people contributed to it, the 469-page report was ultimately reduced to the opinion of two people. As the final draft was completed in mid-September of 1964, Goldberg was essentially given one week to polish it, an impossible task. Time pressure, the rush to publish before the national elections, had bedeviled the commission workers and pushed them into a final homogenized result that could not have pleased anyone but its two authors.

The third failing of the commission was that it had a potentially contradictory agenda. One goal was to ascertain the facts while another was to ensure the nation's reputation and by inference the reputation of its agencies. As Epstein pointed out, having multiple goals was a recipe for problems if those goals ever became mutually exclusive. This created a self-imposed minefield for the commissioners to navigate. While the junior staff pursued the facts and the truth, the commission's own members had decided to join, or had been persuaded to join, the commission to mitigate the potential for damage to America's international image. This created a disconnect between the junior staff performing the investigation and the commission members who ultimately determined the conclusions stated in the final report. Multiple goals led to internal friction and confusion and hampered the commission's efforts to uncover the truth.

The fourth major problem was that the commission decided to rely on government agencies to do the bulk of the investigative fieldwork, supplemented by their junior counsel. Government agencies have their own agendas. They always have to be concerned with their funding from Congress and therefore with their public image. If their image is compromised, loss of funding, reorganization, or relocation could follow. By the 1960s many agencies had become powers unto themselves. Therefore, although the resources of agencies like the FBI, the CIA, and the Secret Service were estimable, if, for any reason—and that reason could include national security as well as maintaining their public image—any of those agencies had a conflict of interest with the investigation, or any linkage that they wanted to hide, the overall success of the investigation would be compromised. This constituted a serious naïveté on the part of the commission and was a potential source of error unrecognized from the outset.

Another serious problem with the commission's investigation was that it denied itself the use of forensic investigative techniques. The staff's role as "independent investigator" was impeded by its role as "commission counsel." Although the commission was authorized to use investigative techniques, it refused to employ them. If the counsel investigators couldn't cross-examine people, couldn't confront witnesses with contradictory testimony or evidence, then they had no means to verify truth. And without the use of those tactics they were limited to just taking depositions and to what the witnesses would reveal openly.[26]

Contrast this with the Watergate investigation in which public hearings were conducted openly and witnesses were cross-examined and confronted with other evidence and testimony. That investigation was ultimately successful at getting at the truth, whereas the Warren Commission failed.

A fifth key problem was time pressure. Rankin had set a deadline of June 1, 1964, for the chapter reports. However, the junior counsel members had refused to begin their separate investigations until Jack Ruby's trial had concluded, for fear of compromising his legal rights. This meant the real investigations did not get started until the middle of March.

This decision was pure folly on the part of the commission attorneys. Fifty million people had seen Ruby shoot Oswald on national television. The facts of the case were not in dispute. Ruby's only defenses were temporary insanity or a "crime of passion" defense that was available to him in Texas in the 1960s. If Ruby could have shown that he acted out of uncontrollable rage or emotional anguish, he may have gotten a sentence as low as two years, or even conceivably probation. But Ruby's state of mind was an issue for him and his defense team. An investigation into the assassination by the Warren Commission could have in no way compromised this type of defense, unless it had turned up evidence that Ruby had co-conspirators. But this was hardly an infringement on Ruby's rights since law enforcement would routinely seek evidence of conspiracy in a crime. Direct personal interviews with Ruby could have waited until after his trial while other aspects of the investigation proceeded. Even if the commission's investigation had somehow compromised Ruby's rights, a strong argument could be made that the national and international importance of the task before the commission superseded the rights of a single cold-blooded killer.

While the deadline of June 1, 1964, could and did slide back, the more rigid political reality of the November presidential election could not be altered. The commission would find it essential to finish its report prior to the election involving the sitting president and political candidate who had commissioned it. This deadline could not easily be moved and would force the writers of the report to reach final conclusions whether they were ready to or not. A clear example of this is provided by Rankin's comment to Wesley Liebeler concerning his July criticisms of the report's conclusions concerning the timing of Oswald's visit to Mexico. Liebeler's analysis left open the possibility that Dallas resident Sylvia Odio's contention that Oswald had come to visit her at her apartment before the assassination was the truth, since the timing did not conflict with the timing of Oswald's trip to Mexico, thereby demanding further investigation. "At this stage," Rankin replied, "we are supposed to be closing doors, not opening them."[27]

But that is the purpose of an investigation: to open doors, to find out what's behind them, and then to report it. It doesn't matter what stage the investigation is in, if something important comes up, it should be followed through. This is the only way to reach the truth. Unfortunately for the commission, time pressure problems meant that many important issues were just left unresolved.

Another severe problem for the Warren Commission was that it disbanded after releasing its report. This meant that it could not respond to criticism. It could therefore not address the substantive issues later raised by its critics.

In science, publication is the beginning of the process, not the end. When a scientist publishes a result, he or she hopes that it will generate enough interest for other scientists to attempt to reproduce the work and for theoreticians to attempt to build models relying on it. Scientists want and expect feedback from their efforts. They fully expect to address that feedback at scientific conferences or in print. The worst thing is for a publication to be ignored. Scientists take citations very seriously—that is, other researchers referencing their works—and they always hope that one of their publications will become a "citation classic," referenced hundreds or thousands of times by scientists around the globe.

But the commission had no hope of addressing feedback after release of its report because it no longer existed as an entity. This was a debilitating failure and one from which it was impossible to recover. In the absence of the original commission, critics multiplied exponentially over the years. It was left to individual workers like David Belin to attempt to defend single-handedly the various reports' problems in print.

The single most important failing of the Warren Commission, however, the one unforgivable act by its members, was the alteration of data. Gerald Ford, the commission's biggest proponent of the single-bullet theory, decided to move the wound in the president's back to one in his neck so that it could account for the throat wound and support the single-bullet theory. This was inexcusable.

The first draft of the Warren Commission report stated that "a bullet had entered his back at a point slightly above the shoulder and to the right of the spine." A bullet fired at a downward angle from the height of the sixth-floor window could not have exited through Kennedy's Adam's apple. Commissioner Gerald Ford therefore deliberately changed the draft to read: "A bullet had entered the base of the back of his neck directly to the right of the spine." When this became public in 1997, due to efforts of the Assassination Records Review Board, Ford responded that he had made the changes for "the sake of clarity."[28]

Changing, fabricating, or manipulating data is anathema in the scientific community; the disapprobation for this activity is enormous in science because data are the stock in trade of the scientist. Scientists caught doing this would suffer grave setbacks to their reputations and careers. Altering data is a sure way to reach the wrong hypothesis and to be ruined professionally. But it can be a way to secure funding in a competitive area of research where scientists are rewarded for producing novel and significant results. Scientists who alter data do so because they have another agenda other than merely uncovering truth. The commission's decision to change the back wound to a neck wound precluded the possibility of it formulating the correct model for the assassination, and is a telltale indication that truth seeking was not in fact the primary goal of the commission.

By far the most serious error of the Warren Commission was altering the data to fit a preconceived model. This activity is not acceptable in any domain of human inquiry, neither law nor science nor religion, and is a guaranteed way to get the wrong answer. Edward Epstein's comment on this point is a remarkable example of veiled polemic, excoriating in its degree of understatement:

> If the FBI reports are accurate, as all the evidence indicates they are, then a central aspect of the autopsy was changed more than two months after the autopsy examination, and the autopsy report published in the Warren Report was not the original one. If this is in fact the case, the significance of this alteration of facts goes far beyond merely indicating that it was not physically possible for a lone assassin to have accomplished the assassination. It indicates that the conclusions of the Warren Commission must be viewed as an expression of political truth.[29]

Because of its academic imprimatur, Edward Jay Epstein's book *Inquest* was taken very seriously by the press, the public, and the university community. It played a crucial role in casting doubt on the conclusions of the commission, and ultimately helped to lead to a reopening of the investigation into Kennedy's death by the House Select Committee on Assassinations in 1978.

CHAPTER 3

THE WITNESSES

Fifty-one witnesses, gentlemen of the jury, thought they heard shots coming from the grassy knoll, which is to the right and front of the president.

—Kevin Costner as Jim Garrison in Oliver Stone's film *JFK*

He who has ears to hear, let him hear.

—Mark 4:9

Any lawyer will tell you that eyewitness testimony is the least reliable evidence you can have. Witnesses often lie, distort the truth, forget, superimpose their own values, become confused, and fabricate on demand. But with physical evidence to support it, eyewitness testimony can illuminate crucial details to an event. In science, if a witness says a rock crashed through her roof and landed on her living room sofa, it seems at first incredible. But if she then produces the rock and it turns out to be a meteorite, then the physical evidence supports her account and her statement is assigned credibility.

In practice, some areas of science depend heavily on the eyewitness testimony of amateurs. Many valuable fossils are found in quarries. New asteroids are often discovered by amateur astronomers. Most scientists working in the field of meteorites have never seen a meteor hit the ground from space. They rely on amateurs to uncover anomalous-looking rocks in places like deserts and ice floes and bring them to professional researchers. The coelacanth, a living fossil thought extinct for 300 million years, was brought to the attention of paleontology by local fisherman in the Comoro Islands in the Indian Ocean.

HEAD SHOT

Science advances by taking the eyewitness testimony of amateurs seriously enough to investigate further. There is a leap of faith when a professional astronomer devotes telescope time to investigate a potential sighting by an amateur astronomer. There is a leap of faith when a paleontologist takes her time to examine a fossil found by workers digging in a quarry or by native tribesmen scouring the landscapes for ancient human bones in the rift valley systems of Africa. But amateurs can and do make substantial contributions to astronomy and paleontology. Of course, not all such findings pay off in terms of discovering something scientifically new or valuable, but enough do that science advances as a result. Some of the very best and most illuminating anthropological fossils have been found by amateurs. The recent discovery of the 47-million-year-old lemurlike fossil, nicknamed Ida, was uncovered in the Messel fossil pit in Germany in 1983 but didn't reach the eyes of the scientific community until it was purchased from a fossil dealer by Dr. Jorn Hurum of the University of Oslo in 2009.[1] Some scientists believe this fossil significantly aids our understanding of the evolution of primates.

Modern models of human evolution depend heavily on crucial details of form and function gleaned from these rare and precious fossils. So it pays to investigate. Eyewitness testimony is very valuable in science. Not all such testimony is reliable, but enough is to make its pursuit a worthwhile endeavor. The advancement of knowledge depends on it. But eyewitness testimony only has value in science— only has meaning—when it is supported by physical evidence.

Immediately after Kennedy was assassinated, and over the years since, numerous witnesses to the assassination have reported seeing or hearing shots fired from the grassy knoll in Dealey Plaza. Some estimates account over fifty witnesses reporting to have heard shots or seen smoke emanating from that grassy knoll. There are photos taken moments after the shooting showing people racing toward the knoll in a bold effort to find the assassin.[2] These people must have been full of adrenaline, clouding their judgment, because anyone who had just killed Kennedy with a rifle could easily do the same thing to them.

But without physical evidence to support them, these people were silenced. Their voices were not heard, and they were soon forgotten. The Warren Commission heard testimony from some of these witnesses but did not believe them. No guns were found on the grassy knoll. No shell casings were found there. Some footprints were found behind the picket fence, but these hardly constitute unambiguous evidence of an assassin. But if an assassin did fire from this position and, unlike Oswald, made his escape and took his rifle and his shells with him, the best, most unmistakable evidence of his presence would not be found. Absence of evidence does not constitute evidence of absence. Even if the proverbial smoking gun was never located, it doesn't preclude the possibility of the existence of other evidence of the assassin's presence. The testimony of so many witnesses attesting to a shooter on the grassy knoll merits a search for evidence to support the assertion. Much in the same way an astronomer would attempt to track an earth-crossing asteroid reported by an amateur with a telescope in his backyard, a thorough scientist would conduct a search for some trace, remnant, or physical evidence of an assassin firing from the grassy knoll.

Detailed recollections by many of these witnesses are reported in Jim Marrs's excellent work *Crossfire: The Plot That Killed Kennedy* (1989).[3] Marrs is a freelance author and photographer who taught a course on assassinations at the University of Texas. Figure 4 shows a diagram of Dealey Plaza and the locations of some of the witnesses at the time of the assassination.

One of the first witnesses to report hearing a shot from the grassy knoll was Bill Newman, who was situated on Elm Street just below the concrete cupola. He recalled that "a shot took the right side of his [Kennedy's] head off. His ear flew off. I heard Mrs. Kennedy say, 'Oh my God, no, they shot Jack!' He was knocked violently back against the seat, almost as if he had been hit by a baseball bat. At that time I was looking right at the president and I thought the shots were coming from directly behind us."[4] Cheryl McKinnon, a journalist who was standing next to Bill Newman, recalled, "Suddenly three shots rang out in succession. Myself and dozens of others standing nearby turned

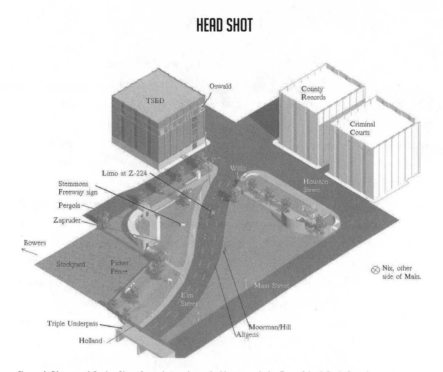

Figure 4: Diagram of Dealey Plaza from the southeast, looking toward the Texas School Book Depository with the limo at the Z224 position. Witnesses and fixed objects noted.
Image © 1996 Paul J. Burke. Reprinted by permission of Paul J. Burke.

in horror toward the back of the grassy knoll where it seemed the shots had originated. Puffs of white smoke still hung in the air in small patches."[5]

Emmett J. Hudson, a Dealey Plaza groundskeeper, was sitting on the steps leading to the top of the grassy knoll. He recalled that "when that third shot rang out... you could tell the shot was coming from above and kind of behind."[6] Based on Hudson's position, this would have been the location of the picket fence on the grassy knoll.

Jean Hill, who was standing next to Mary Moorman when she took her famous Polaroid photograph of Kennedy moments after he had been shot in the head, had a clear view of the fatal shot. She claimed to have seen a man fire from behind the wooden picket fence on the grassy knoll and saw smoke drift from this location. She watched in horror as the man fled from the scene.[7] This account was corroborated by the testimony of a railroad supervisor, S. M. Holland, who told government investigators that he also observed smoke drifting from the knoll: "A puff of smoke came out about 6 or 8 feet above the ground right out from under those trees. And at just about this location from where I was standing you could see that puff of smoke."[8]

Mary Woodward, who worked for the *Dallas Morning News*, was standing in front and to the left of the wooden fence. She would later write that "suddenly there was a horrible, ear-shattering noise coming from behind us and a little to the right."[9] Abraham Zapruder, who was taking movie pictures of the assassination from a concrete cupola situated to the east of the knoll, told a Secret Service agent that he thought the shots came from behind him.[10]

One of the most interesting accounts is from a deaf man named Ed Hoffman of Dallas, who told a clear story of seeing a man carrying a rifle running west along the back side of the wooden fence overlooking Dealey Plaza. He stopped and tossed his rifle to a confederate who broke it down and placed it into a railroad bag immediately after the assassination. Hoffman watched as both men fled the area. Moments later, he saw Kennedy's motorcade come into sight from under the Triple Underpass. The president was lying down in the blood-splattered limousine. Only then did Hoffman realize what had happened. He witnessed these events while walking near the railroad tracks behind the grassy knoll area. Partly because of his disability, it took several years for his story to reach the authorities.[11]

In the book *Ultimate Sacrifice*, assassination researcher Lamar Waldron also reports accounts of multiple witnesses recalling events consistent with a shooter on the grassy knoll. Lee Bowers, a railway supervisor who was located in a tower in the railroad yard behind the knoll, recollected that a series of cars drove into the area and left between 11:55 and 12:25 p.m. One of the drivers appeared to have a microphone or walkie-talkie device with him. Two of the cars, a 1959 Oldsmobile and a 1961 Chevrolet Impala, had "Goldwater for 65" bumper stickers.[12] By 12:28 p.m., Mr. Bowers noticed two men behind the stockade fence, looking toward Main and Houston streets. One man he described as middle-aged and heavyset, wearing a white shirt and dark trousers. The other man was in his midtwenties and stood about ten to fifteen feet from the first man. He wore a plaid shirt or plaid coat.[13]

Gordon Arnold, a soldier, claims to have been walking behind the picket fence on the grassy knoll when he was confronted by a man who showed him a badge, told him he was a Secret Service agent, and asked him to leave the area. Arnold then moved to the other side of the fence on the grassy side between the fence and the street. The Secret Service has always maintained that no agents were present on the grassy knoll or anywhere on the ground in Dealey Plaza prior to the assassination, as detailed in the files of the Warren Commission.[14] When the shots rang out, Arnold hit the ground in self-defense. He would later say, "The shot came from behind me, only inches over my left shoulder. I had just gotten out of basic training...and I hit the dirt."[15]

Riding two cars behind the president was Texas senator Ralph Yarborough, a World War II combat veteran. He confirmed the basics of this account: "Immediately on the firing of the first shot I saw the man ...throw himself on the ground...he was down within a second, and I thought to myself, 'There's a combat veteran who knows how to act when weapons start firing.'" However, Yarborough's description may have referred to Bill Newman, who was lying on the ground after the shooting, together with this wife, near the street. Additionally, Secret Service agent Lem Johns would later testify, "The first two [shots] sounded like they were on the side of me toward the grassy knoll."[16]

As shots were fired at Kennedy's motorcade, Lee Bowers claimed to have seen from his vantage point in the tower a flash of "light or smoke" from the two men behind the fence.[17] He thought he recalled seeing one of the men standing on the bumper of a car backed up to the fence. After the shots ended, a railroad yardman recalled seeing someone behind the fence on the knoll "throw something in a bush." J. C. Price, a building engineer who was standing on the roof of the Terminal Annex Building at the south end of Dealey Plaza, across from the grassy knoll, claims to have seen from his vantage point a man running full speed away from the fence toward the railroad yard, carrying something in his hand that "could have been a gun."[18] In an affidavit signed that day, Price further asserted: "There was a volley of shots, I think five." He went on to say that the shots came from "just behind the picket fence where it joins the underpass."[19]

Dallas sheriff Bill Decker, who was riding in the front car of the motorcade, radioed for all available men to converge on the railroad yard behind the fence on the grassy knoll. Photographs, taken shortly after the assassination, show a policeman running toward the knoll. The Newmans can be seen lying on the ground. However, Gordon Arnold is not seen. Unless he is between the fence and the concrete cupola, or he quickly got up and ran, this picture fails to provide confirmation to his story. However, Mr. Arnold's account aside, of the twenty deputies who provided statements, sixteen of them thought that "the assassin had fired from the area of the grassy knoll."[20]

Dallas Police Officer Joe Smith claimed to have smelled smoke from gunpowder when he arrived behind the fence with his deputy. As

Figure 5: Policeman running toward the grassy knoll. Taken by Wilma Bond after the assassination.

he moved into the area, he "caught the smell of gunpowder there," later recounting, "I could tell it was in the air."[21] He recalled drawing his revolver and confronting a man who then claimed to be a Secret Service agent despite being dressed only in a sports shirt and sports pants instead of the typical dark suit. Officer Smith then allowed the man to leave the area. This account was confirmed by Sheriff's Deputy Seymour Weitzman, who also told the Warren Commission that he saw a "Secret Service agent" behind the fence. Weitzman went on to say that he found footprints behind the fence that appeared to be going in "different directions." Officer Smith searched a 1960 Chevrolet Sedan parked by the fence, but did not look in the trunk.[22]

These officers were shortly joined by three railroad employees who had witnessed the assassination from the Dealey Plaza overpass. These men had heard shots from the grassy knoll and had seen a cloud of smoke hanging over the trees there. The railroad men not only claimed to have seen the footprints behind the fence, but also a pair of muddy footprints on the bumper of a station wagon parked there.

An off-duty policeman, Officer John Tilson, recalled seeing a man "slipping and sliding" down the railway embankment behind the knoll. He recounted that the man ran to a parked black car, "threw something in the back seat," got in the front seat, and sped off. Tilson tried to follow the car but lost it in traffic.[23]

Of the railroad workers standing on the overpass, seven reported seeing smoke emanate from above the bushes on the grassy knoll.[24] Six other men from the overpass also reported seeing smoke in this area. Austin L. Miller provided an affidavit in which he certified, "I saw something which I thought was smoke or steam coming from a group of trees north of Elm off the railroad tracks."[25]

Additional witnesses also recounted seeing smoke near the bushes and trees near the corner of the wooden picket fence. James L. Simmons recounted that the sounds of gunshots "came from the left and in front of us, toward the wooden fence, and there was a puff of smoke that came underneath the trees on the embankment." Richard C. Dodd also recalled that "the smoke came from behind the hedge on the north side of the plaza."[26] Several other eyewitnesses in the vicinity also reported seeing smoke or "white smoke" in the trees near the knoll.

A journalist who had witnessed the assassination, Robert McNeil, would later write that he saw "a crowd, including reporters, converged on the grassy knoll believing it to be the direction from which the shots that struck the president were fired."[27] McNeil himself would follow this crowd, proof perhaps of a herding mentality in Dealey Plaza as bystanders hoped to somehow "catch" Kennedy's assassin. It certainly can't be ruled out, however, that these people were responding en masse to their own individual perceptions.

Not only did people who were near the grassy knoll at the time of the shooting report shots coming from the railroad yard area, but some witnesses who were at or inside the Texas School Book Depository during the assassination also reported hearing shots fired from this westerly direction. Ochus B. Campbell, the book depository vice president, who was standing in front of the depository at the time of the motorcade's passing, recalled, "I heard shots being fired from a point which I thought was near the railroad tracks located over the viaduct on Elm Street." Otis N. Williams was also standing on the steps in front of the building. He told FBI agents, "I thought these blasts or shots came from the direction of the viaduct which crossed Elm Street."[28] Danny Garcia Arce was standing on the grass in front of the Texas School Book Depository. He recounted that he thought the shots "came from the direction of the railroad tracks near the parking lot at the west end of the depository Building."[29]

Victoria Elizabeth Adams was on the fourth floor of the Texas School Book Depository, watching the motorcade through a window. After the last shot was fired, Ms. Adams and a coworker "ran out of the building via the stairs and went in the direction of the railroad where we had observed other people running."[30] Although unasked, she nonetheless offered her opinion as to where the shots originated to the commission's counsel: "It seemed as if it came from the right below rather than from the left above."[31] Dorothy Anne Garner also witnessed the motorcade from the fourth floor of the depository. She recalled that she "thought at the time the shots or reports came from a point to the west of the building."[32] This would have been toward the grassy knoll area and the railroad tracks.

Two Secret Service agents riding in the motorcade also thought that shots originated from the grassy knoll. Agent Forrest V. Sorrels was riding five car-lengths in front of Kennedy. At the sound of shots, Sorrels looked immediately toward the knoll on his right. "The noise from the shots sounded like they may have come back up on the terrace there...the reports seemed to be so loud, that it sounded like to me— in other words, that was my first thought, somebody up on the terrace, and is the reason I looked there."[33] Agent Paul E. Landis Jr. was standing on the right running board of the car following Kennedy's limousine. He recounted, "My reaction at this time was that the shot came from somewhere toward the front."[34]

Two of the most thoroughly credible of the grassy knoll witnesses, however, were presidential aides Kenneth Phillip O'Donnell and David Francis Powers. Both men knew the Kennedys well and had worked closely with them for years.

O'Donnell was born in Worchester, Massachusetts, in March 1924. During World War II, he served in the Army Air Corps, matriculating to Harvard after the war. It was there that he met Robert Kennedy. He later went on to Boston College Law School. As a member of the Democratic Party, O'Donnell helped John F. Kennedy get elected to the US Senate in 1952. In 1957 he was appointed assistant counsel to the Senate Select Committee to Investigate Improper Activities in Labor-Management Relations.

By 1960 O'Donnell had become the director of John F. Kennedy's presidential campaign schedule, eventually becoming Kennedy's special assistant and appointments secretary. However, he was essentially Kennedy's chief of staff, a position that the president never officially filled. O'Donnell was one of the first critics of the Vietnam War and had urged Kennedy to terminate America's involvement in Southeast Asia.

David Francis Powers was born in Massachusetts in 1912. After finishing school, he worked for a publishing company and attended Boston

THE WITNESSES

University at night. During World War II, Powers joined the air force and served with the Flying Tigers in Japan, China, and Burma. Powers met Kennedy during his 1946 campaign. Working primarily in the Charleston area of Massachusetts, his hometown, Powers helped Kennedy win the 11th Congressional District. Powers also helped Kennedy win the Senate seat against Henry Cabot Lodge in 1952. After Kennedy became president, he appointed Powers as his special assistant. In that capacity, Powers became one of Kennedy's closest advisers. Kenny O'Donnell once said, "Outside of Bobby, President Kennedy had one really close friend and that was Dave Powers."

Both Kenny O'Donnell and Dave Powers were riding in the car behind Kennedy in the Dealey Plaza motorcade. They were in a perfect position to ascertain the source of the shots, because they had a clear view of both Kennedy and the grassy knoll. House Speaker Tip O'Neill revealed in his autobiography that, five years after the assassination:

> I was surprised to hear O'Donnell say that he was sure he had heard two shots that came from behind the fence.
>
> "That's not what you told the Warren Commission," I said.
>
> "You're right," he replied. "I told the FBI what I had heard, but they said it couldn't have happened that way and that I must have been imagining things. So I testified the way they wanted me to. I just didn't want to stir up any more pain and trouble for the family."
>
> Dave Powers was with us at dinner that night, and his recollection of the shots was the same as O'Donnell's.[35]

While it would be a mistake to argue that all the witnesses who claimed to have seen smoke or a shooter, or heard shots from behind the grassy knoll, are credible, with so many witnesses, it would be equally mistaken to say that they are all lying or hallucinating. This is an especially difficult hypothesis to accept when many of these accounts are consistent and corroborate each other. Even if each witness has only a one-in-ten chance of being correct, the odds that all fifty are mistaken are about five chances in a thousand. Therefore, assigning a low-ball probability of accuracy to each witness still yields odds greater than 99 percent that at least one of them provided a true account of the events that transpired in Dealey Plaza, and therefore that an assassin fired at Kennedy from the grassy knoll. Dismissing all of these witnesses out of hand and failing to investigate further is therefore an egregious error in judgment.

But in science, eyewitness testimony only has meaning when supported by physical evidence. The question then remains, can substantial empirical evidence be found to support the contention that one or more shots were fired from the grassy knoll and provide these myriad accounts with meaningful scientific value and logically consistent credibility?

CHAPTER 4

HOW SCIENCE ARRIVES AT THE TRUTH

Those who say that the sun does not move and that it is the earth that revolves and that it turns, pervert the course of nature.
    —John Calvin, sixteenth-century French theologian

I do not feel obliged to believe that the same God who has endowed us with sense, reason, and intellect has intended us to forgo their use.
    —Galileo Galilei

Joseph Wiener knocked hard on the heavy wooden door of Charles Dawson's neighbor's home. Finally it opened; an old English gentleman peered warily at him.

"I wonder if I might talk with you about your neighbor of long ago, a Mr. Dawson," he said.

"You want to talk to me about Mr. Dawson, about Piltdown Man?"

"Yes," he said, "that's exactly what I want to talk about."

"Why? Do you think perhaps there's something wrong with it?"

"Yes. As a matter of fact, I think there's something very wrong with it. I think it's a forgery."

"Well do come in my good man," the English gentleman replied, "I've been expecting you for forty years."

    —Joseph Weiner, *The Piltdown Forgery* (1955)[1]

Piltdown Man was the major scientific hoax of the twentieth century. It fooled a generation of anthropologists, wasted dozens of scientific careers, and hindered progress in the field of anthropology for decades. The hoax was perpetrated by Charles Dawson, an Englishman, perhaps in collaboration with Pierre Teilhard de Chardin, a Frenchman. It succeeded spectacularly in part because it played into scientists' preconceived notions of how they thought human evolution should have occurred and because it stoked the national pride of the British scientific community. The hoax was eventually uncovered through diligent application of the scientific method.

In 1909, Charles Dawson, an amateur fossil collector, unearthed a fossil skull in a shallow gravel pit in a quarry in Sussex, England, called Piltdown. The skull appeared to be human, but by 1912, a very primitive matching jawbone was also found in the same area. Fossils from Pleistocene fauna had also been found in the quarry, including elephants, mastodons, and rhinoceroses. These fossils proved the geological era of the quarry to be millions of years old. By September 1912, enough evidence had accumulated for Dawson, together with Arthur Smith Woodward of the British Museum of Natural History, to announce to the Geological Society of London the discovery of an ancient human fossil they had assigned the scientific name of *Eoanthropus dawsoni*. The fossil was commonly called Sussex Man, Dawn Man, and finally Piltdown Man. By 1915, Dawson had discovered a second fossil skull that corroborated the original find. This fossil was dubbed Piltdown II.

Although Piltdown Man had a skull that appeared modern in size and characteristics, the accompanying jawbone was very primitive. It was so primitive, it appeared simian. Only a half century before, Charles Darwin had announced his theory of evolution to the world. In a follow-up to the 1859 publication of *On the Origin of Species*, Darwin published *The Descent of Man* in 1870, in which he argued that humans were descended from apes, a view that was anything but universally embraced, and in some quarters, still controversial even today. However, the species that represented a transitional creature between humans and apes had yet to be found.

Fossils of a humanlike species had been discovered in the Neander Valley in Germany in the 1850s, and fossil remains of Cro-Magnon

Man, our forbearers, had been found in France in the late 1800s. Even more primitive fossils had been unearthed in Borneo, such as "Java Man" in the 1890s. These species appeared to be very humanlike. They clearly walked erect and had many features in common with modern humans. However, the "missing link," a hypothetical creature that represented a species halfway between apes and man, the ancestor of modern humans, had been predicted to exist on theoretical grounds but had yet to be discovered.

In one fell swoop, Piltdown Man solved a series of problems for the anthropological community and in particular for the British scientific community, whose members had a long and illustrious place in the history of science starting with the work of Robert Hooke and Isaac Newton and the formation of the Royal Society. However, at that time, all the major finds of early humans were being made in Europe and Southeast Asia, while the British Isles were not even on the anthropological map.

Piltdown Man changed all that. It provided Britain with a fossil of even greater antiquity than any that had been found so far. The features and age of the Piltdown fossils indicated that they were very primitive and of ancient vintage. Nothing found in Europe or Asia could compare with it. It was clearly of tremendous paleontological importance and provided crucial details of the early evolution of man. England had leapt from the role of anthropological afterthought to world leader in the field in a geological heartbeat.

Piltdown Man told the anthropological community something very important about human evolution. It seemed to fit in well with theories of the "missing link" that had been proposed. Best of all, it exhibited the traits that some scientists had predicted on evolutionary grounds. Many believed that the evolution of a large brain was the first development that separated us from our simian forbearers. It was only after the institutions of early civilization had become established due to early man's intellectual prowess that the development of other modern traits began. Thus, the recession of large canine teeth, characteristic of all apes, into small cuspids, which are a distinguishing feature of humanity, took place as the advent of proto-civilization no longer made threatening fangs necessary. Many believed and had predicted that changes in jaw structure and vocal apparatus took place as a result of higher intelligence, not as a prerequisite for it, nor as a trait that developed collaterally.

In 1924, the Australian physical anthropologist Raymond Dart announced the discovery of a small primitive skull in the Taung region of northwestern South Africa. The skull was that of a child, but its characteristics were unlike either chimpanzees or humans. The brain case was small, about the size of a chimp's at a comparable age; however, the dentition, the form and arrangement of teeth in the mouth, appeared modern. The position of the spinal cord as it entered the skull suggested that the creature walked erect. The age of the fossil was dated to strata two and a half million years old. Dart named the creature *Australopithecus africanus*, which means "southern ape from Africa."

However, Dart's announcement and his interpretation of the fossil met with great skepticism from the scientific community. Many believed it was just another species of chimpanzee and exhibited no advanced features like bipedal locomotion. Others believed that Dart's conclusions required "mosaic" evolution, the development of some features while others remained stagnant, a view that challenged their preconceptions of how Darwinian evolution should occur. Still others thought that since Dart's fossil was that of a six-year-old or younger child and the features of children are known to differ from those of adults due to neotony (the appearance of traits in a youngster that are not as dramatic and pronounced as those in an adult), it was premature to make evolutionary conclusions until an adult fossil was found. Further, Dart's view that his Taung child exhibited advanced human traits like bipedalism contrasted sharply with the theories of anthropologists Sir Arthur Keith and Elliot Smith, who argued that hominization, the evolutionary process of becoming human, began with an enlarged cranial capacity. Keith and Smith's conclusions, of course, were based on the Piltdown fossils. For these reasons, many felt that Dart's conclusions of a new species that was protohuman were unjustified. The scientific community preferred to classify Dart's discovery as a kind of ancient ape.

HOW SCIENCE ARRIVES AT THE TRUTH

Even though Darwin himself had predicted that ancestral forms of humans would likely be found in Africa, basing this view on his conclusion that humans were most closely related to African chimpanzees, the contrast between this fossil and the well-established Piltdown finds could not be ignored. It was easier to place Dart's new species in with the chimps rather than to assign it to the human lineage. However, as more human fossils were found in places like Java, China, and South Africa, they exhibited different characteristics from Piltdown Man. Like the Taung skull, their brain cases were far more apelike than the Piltdown fossils and their jaws much less so. They formed a consistent picture of evolutionary lineage that did not agree with Piltdown Man. Defenders of the fossil argued that the apparent extreme age of Piltdown Man indicated that two separate lines of hominids might have arisen contemporaneously. In this view, there were two distinct lines of human evolution: one arising from Piltdown Man and the other from a different ancestral type.

However, as even more ancient fossils began to accumulate it was becoming apparent to most anthropologists that Piltdown Man was an anomaly. No other fossil discoveries were at all similar to it. Through the 1930s and '40s, serious doubts as to the validity of Piltdown Man were beginning to develop within the scientific community. In 1929, the first skull of Peking Man was discovered in China. New australopithecine fossils appeared in Africa together with additional *Homo erectus* finds. A clear and undeniable pattern was emerging. In his book *Bones of Contention*, Roger Lewin, a distinguished British anthropologist and science writer, quotes anthropologist Sherwood Washburn: "I remember writing a paper on human evolution in 1944, and I simply left Piltdown out. You could make sense of human evolution if you didn't try to put Piltdown into it."[2]

In 1948, *The Earliest Englishman* by Arthur Smith Woodward was published by Watts and Co. London. This was the last work to appear in print that took Piltdown Man seriously as a fossil and drew anthropological conclusions from it. This work appeared almost forty years after the original find in the quarry.[3]

HEAD SHOT

By 1949, however, the newly developed technique of fluorine dating was applied to the Piltdown fossils. This technique is based on the observation that bones absorb fluorine from the soil. Over time, this element accumulates in a fossil to a degree dependent on the length of time the fossil was in the soil. To the surprise of many, the Piltdown fossils showed very little presence of fluorine, while other fossils "discovered" in the Piltdown gravel exhibited relatively high fluorine concentrations. This technique dated both the Piltdown skulls and the jaw to a period less than 50,000 years ago. Now things had become complicated. Since Piltdown Man was not of great antiquity, as many in the anthropological community had assumed, its primitive dentition and jawbone could not easily be reconciled with extant fossil records of other hominids. Piltdown Man was therefore an unsalvageable anomaly.

In 1953, Oxford anthropologists Joseph Weiner and Le Gros Clark together with paleontologist Kenneth Oakley announced the solution to the problem of Piltdown Man's anomalous features: it was a forgery. They had made this determination from careful examination of the original fossils themselves. On these fossils was clear evidence of chemical staining, applied to make the fossils appear to be of ancient lineage. Second, there were clear indications of filing of the teeth and molars, as well as deliberate destruction of the mandible at the point that it fit into the skull. Since the skull and jawbone did not match up, destruction of the region of attachment prevented anthropologists from readily observing that the fossils were from two different species. Not only had these fossils been fakes, but the Piltdown quarry had been "salted" with other animal fossils, primarily teeth, to make it appear that the strata were of Pleistocene or Pliocene age. After Oakley and Weiner had presented their results to the Geological Society of London in June 1954, Sir Gavin de Beer, director of the British Museum (Natural History), concluded a broadcast with these words: "We have laid the ghost of Piltdown Man to rest, who, as it happens, never fitted very happily into any scheme of man's evolution. Indeed we have gained, because half a dozen new experimental methods of studying fossils have now been developed which will not only make any repetition of such a hoax impossible in future, but will materially assist the scientific study of fossils."[4]

Ultimately, the Piltdown fossils were carbon-dated, confirming that the human skull was only six hundred years old, while the orangutan jaw was about eight hundred years old. Fossils from the two distinct species, of recent origin, had been cobbled together to suggest the appearance of a new human ancestor in the fossil record.

Suspicion immediately fell on Charles Dawson, a professional solicitor but an amateur paleontologist, who was the principal discoverer of the fossils. Suspicion also attached to the codiscoverer, Pierre Teilhard de Chardin, who was a student at the time. Few believed that Woodward of the London Museum had any involvement in the forgery. Joseph Weiner, however, was firmly convinced that Dawson was in fact the perpetrator.

One of the problems that hindered the timely discovery of the hoax was the way the fossils were presented for inspection. The Piltdown fossils were considered so rare and so valuable that any researcher who wanted to study them had to request permission in advance. When they came to examine the fossils, believed to be of great antiquity and importance, the fossils were brought out from the safe together with plaster casts that had been made of them. The visiting anthropologist was allowed to look at the original fossils and the casts briefly to see how perfect the plaster casts modeled the original bones. When the anthropologist was convinced of the accuracy of the models, the original fossils were placed back in their safe, and the casts were left with the researcher to examine and study. Measurements and assessment of the features of the fossils were therefore made strictly through examination of the casts, not the original fossils themselves. This made it impossible to determine that the original bones had been filed down and stained to make them appear ancient. This determination, of course, could not be made from the casts, which accurately reproduced the dimensions of the hoaxed fossils but not the evidence of forgery.

Although Weiner had successfully debunked Piltdown Man and tarnished the reputation of Dawson, he was somewhat circumspect in his book of leveling an accusation of outright fraud: "our verdict… must rest on suspicion and not proof. In the circumstances, can we withhold from Dawson the one alternative possibility, remote though it seems, … that he might, after all, have been implicated in a 'joke,' perhaps not his own, which went too far?"[5]

After Dr. Weiner gave a presentation at the Johns Hopkins University Applied Physics Laboratory in Laurel, Maryland, in 1980 on his work to uncover the Piltdown hoax, I asked him point-blank after his lecture whether he thought that Dawson was deliberately trying to expose the scientific gullibility that pervaded the era, and so whether he was trying to make a point regarding scientists' overeagerness to embrace any idea that supported their own theories. But Weiner replied in the negative. He told me that he didn't think that was Dawson's reason at all, because he had many opportunities to "blow it," to use his vernacular. Because Dawson kept the secret his entire life, Weiner believed that his motivation was to gain fame for himself, rather than to teach the scientific community a painful lesson.

Others, however, have embraced Weiner's idea of a "joke gone too far." Stephen Jay Gould, in *The Panda's Thumb*, has suggested strongly that Pierre Teilhard de Chardin was the mastermind behind the hoax.[6] Inasmuch as Dawson was an amateur fossil hunter, he lacked the knowledge and expertise to pull off the hoax. Chardin had the knowledge and may have thought it fun to play a student prank. When everyone began to take the find seriously, Dawson basked in the limelight, while Chardin attempted to recede back into the woodwork. Things quickly got out of hand for Chardin, and he decided to maintain his silence in order to salvage his career. Some have even suggested that Sir Arthur Conan Doyle, author of the Sherlock Holmes stories, played a role in the forgery.[7] It may be that we will never know the truth. All we know for sure is that Piltdown Man was a fake, a series of separate fossils of modern vintage cobbled together to appear to be something that it was not, evidence of early man in Britain.

Piltdown Man is important for two reasons. First, it demonstrates why scientists never accept eyewitness testimony by itself; it must always be supported by either physical evidence or a reproducible experiment. The greatest scientists of the day accepted Piltdown Man hook, line, and sinker. It wasn't until other evidence came to the forefront that anthropologists began to doubt it. If science had accepted their opinions without evidence to back it up, the fraud could never have been exposed. We would always have to go by what the original scientists thought, even if they were badly mistaken. If the scientific reexamination of the original fossil had not been possible by later generations, the Piltdown fraud would still be plaguing the science of anthropology today. Second, it shows how science ultimately gets to the truth; everything is open to question all the time. When new scientific techniques become available, they are applied to old evidence. Sometimes the new techniques clarify and make determinations more precise. Carbon-14 dating can give very precise dates for the antiquity of a fossil, for instance, when in the past the age of bones was determined by the known age of other samples found in the strata and through geological estimates. Sometimes, however, as in the case of Piltdown Man, the new techniques can expose an outright fraud.

Scientists seek controlled, reproducible experiments or robust physical evidence because conclusions in science are open to question all the time. Nothing in science is sacrosanct. Scientists today are still trying to design experiments to provide more and more rigorous and accurate tests of General Relativity (GR), which is one of the very best theories we have. So far, GR has proven consistent with every experiment and every astronomical observation ever made. Scientists began to suspect a problem with Piltdown Man when other ancient human fossils came to light showing diametrically opposite traits. It was this inconsistency with other fossil data and the development of new paleontological dating techniques that led to Piltdown Man's unmasking. Consistency with other evidence is very important to scientists. If inconsistencies exist in a data set, or between data and a theory, they must be resolved. This is the most important difference between science and pseudoscience; in science, discrepancies are actively investigated. Although Piltdown Man set back and hindered the development of the field of physical anthropology, ultimately, the scientific method prevailed and led to the truth.

Piltdown Man is an old example of how science arrives at the truth. A more modern example of a different kind is provided by the case of a discovery at Brookhaven National Laboratory in the early 1990s concerning the recently announced discovery of cold fusion. On the heels of the controversial announcement that cold fusion had been achieved in a tabletop device by Stanley Pons of the University of Utah and Martin Fleischmann of the University of Southampton, researchers at Brookhaven claimed to have achieved cluster-impact fusion. They had bombarded a titanium-deuterium target with heavy water clusters at an accelerated rate and witnessed an energy release, but most intriguingly, they had done it at room temperature.[8] This seemed extraordinary since extremely high temperatures are typically required to produce fusion reactions with deuterium. This was a much different kind of experiment, however, than that reported by Pons and Fleischmann. Whereas Pons and Fleischmann had claimed reports of "excess heat," very indirect evidence of nuclear fusion reactions produced from their electrochemical experiments using deuterium oxide (heavy water), the Brookhaven researchers were reporting direct evidence, through detection of nuclear reaction products (fast-moving particles) of nuclear processes occurring at astounding rates in an experiment using an accelerator. Because they thought the result was of great significance, they published their discovery of cluster impact fusion in a prestigious journal that specializes in communicating important scientific breakthroughs.

However, while *Physical Review Letters* publishes brief articles outlining new and important discoveries, it also publishes single-page comments, which allow other scientists to respond to its primary articles. Ed Cecil of the Colorado School of Mines published a comment in *PRL* pointing out that the type of accelerator used in the Brookhaven work was a linear one, as opposed to a curved design. In the curved design, it was possible to use a magnetic field to select out the species of particle desired, and so other undesired species in the beam were filtered out. Only the desired species would have the correct charge/mass ratio to make it around the curve. This left open the possibility that pure deuterium ions were being accelerated into their targets. This presence of pure deuterium in their accelerated beam would readily and naturally produce the nuclear reactions they were seeing. The Brookhaven group went back, examined their original findings, and concluded that Cecil was correct. Most of their result had indeed been due to the presence of undetected accelerated deuterium in their beam.

This example, unlike Piltdown Man, is a case of an honest mistake in science. Because other scientists can and do attempt to replicate the work of researchers in other laboratories, and publish their results accordingly, mistakes or errors in an experiment or theory are usually found out very quickly.

With the outstanding exception of Piltdown Man, most hoaxes or errors in science are readily revealed. Another such example is the discovery of N-rays. The famous American physicist R. W. Wood debunked this "discovery" on his trip to French physicist Prosper-René Blondlot's laboratory at the behest of the journal *Nature*. In 1903, Blondlot had announced the discovery of a novel form of radiation he called N-rays. Many scientists had attempted to reproduce his work. R. W. Wood, a professor at Johns Hopkins University in Baltimore, following his own failure to reproduce the result, decided to travel to France to investigate. Wood, who often testified in court cases, was once asked by an attorney during examination, "Who is the best physicist in the world?" He replied, "I am." When a reporter later asked him why he said that, he explained, "Why, my good man, I was under oath."[9]

While at Blondlot's lab witnessing a demonstration of the production of "N-rays" on a screen, Wood surreptitiously removed a prism from the apparatus, an essential component for production of the N-rays, and hid it in his pocket. When he asked if the scientists who were "observing" the N-rays were still seeing them, they replied in the affirmative. Whereby Wood turned on the lights, pulled out the prism, and showed it to them. Wood published his experiences in the September 29, 1904, edition of *Nature*, and the phenomena of N-rays was debunked. Wood wrote:

After spending three hours or more in witnessing various experiments, I am not only unable to report a single observation which appeared to indicate the existence of the rays, but left with a very firm conviction that the few experimenters who have obtained positive results, have been in some way deluded. A somewhat detailed report of the experiments which were shown to me, together with my own observations, may be of interest to the many physicists who have spent days and weeks in fruitless efforts to repeat the remarkable experiments which have been described in the scientific journals of the past year.

Another example of an honest mistake is that of polywater, which was initially believed to be a new and exotic state of water. The Soviet physicist Nikolai Fedyakin had made measurements on the properties of water that had been condensed in, or forced through, narrow capillary tubes. Some of these experiments seemed to indicate the presence of a potential novel form of water with a higher boiling point, lower freezing point, and much higher viscosity than ordinary water, about the consistency of syrup. In 1966, he presented his results in England. Over the next several years, many scientists around the world attempted to reproduce his work. It turned out, however, that polywater was nothing more than ordinary water with some impurities in it. When the experiments were performed with clean glassware, the polywater effects disappeared. The Russian scientists who had announced the discovery were convinced and retracted their initial claims. These types of errors are usually discovered and exposed quickly in science. In the case of polywater, about five years were required to determine the true cause of the anomalous effects. While some have argued that polywater was an example of "pathological science," that is, wishful thinking on the part of the scientist making the claim,[10] in my view it is just science proceeding normally. In some areas of research, particularly in complex biological or chemical systems where minute quantities of impurities can have a significant impact on chemical or physical properties, sorting things out can take time.

HOW SCIENCE ARRIVES AT THE TRUTH

Although the history of science is replete with dead ends, wrong turns, errors, delusions, and hoaxes, ultimately, the scientific method prevails in getting at the truth. If this were not the case, our technology wouldn't work because it is all based on science. It is no accident that radio, transmitted power, and lightbulbs were invented in the nineteenth century, because that is when the laws of electromagnetism were discovered and codified. It is no accident that lasers and computers were invented in the twentieth century, because that is when the underlying theory of quantum mechanics was developed. Our technology is based on science. Science is a method of learning about nature, the practice of which has been fully developed over the last several hundred years.

In fact, science gets to the truth better than any method yet invented. It unerringly uncovers frauds and hoaxes, the worst-case scenarios, as well as honest mistakes. Although this unmasking doesn't always occur immediately, in the end, the method always prevails. This is why a scientific approach to resolving the truth of the JFK assassination has the best possible chance of success. To see why science succeeds and other methods do not, it is enlightening to take a short detour through the history of science.

The very beginnings of protoscience lay in ancient Greece. The Ionian nature philosophers were some of the first to practice a method of natural inquiry similar to that of modern science. The ancient natural philosopher Democritus not only conceived the first version of atomic theory, he had a theory of evolution as well. He believed that the Milky Way was composed of unresolved stars. Not only did the Greeks know that the world is round, they measured it. Eratosthenes used a stick to measure the length of a shadow the sun cast at noon on the summer solstice in Alexandria compared to no shadow when the sun was overhead at noon on the summer solstice at the Tropic of Cancer in Syene, now Aswan, Egypt. He was able to use the length of shadow to determine the angle of the stick and, consequently, the angle of Alexandria relative to Syene along the spherical surface of the earth. Since he already knew the distance between these two locations, this information allowed him to determine the circumference of the earth through simple geometry. Aristarchus knew not only that the earth rotated but that it revolved around the sun as well.

But much of what survived the ancient world, the greatest influences from antiquity, fell to the great natural philosophers of fifth-century Athens. The philosophers of Athens—Socrates, Plato, and Aristotle primarily—led intellectual thought away from the methods of the Ionians to a paradigm where "man was the measure of all things." This de-emphasis of nature and the prominent placement of man led natural philosophy astray to a method of inquiry that stressed pure thought through dialectic reasoning, as opposed to inquiry through observation.

Plato recorded and amplified the arguments of Socrates, who, like Jesus Christ, never actually recorded anything. Socrates used a process called dialectic reasoning, where he attempted to reach the truth through questioning and logical argument. He had a group of followers who apparently hung on his every word, and a coterie of detractors who were often tripped up by him in intellectual arguments, or so Plato tells us. Typically, Socrates would question his critics, lead them into a serious of reasonable admissions, and then show that their cherished beliefs and firmly held views ultimately led to inconsistencies or contradictions. This method of inquiry became known as the Socratic method.

This philosophical approach became so popular in intellectual thought that by the time of the Middle Ages, if people wanted to learn something they just looked it up in the works of Aristotle, a student of Plato. The ancient Athenian philosophers could reason well about humanity, drama, and philosophy, but they almost totally missed the mark when it came to nature. To be fair, some of Aristotle's biological descriptions were not rivaled until the nineteenth century. But when it came to physics, or natural philosophy, as it was called, he was, unfortunately, just plain wrong.

For one thing, Aristotle placed the earth at the center of universe. The earth did not spin, according to Aristotle, because if it did, everyone would just fly off of it. Instead, the heavens rotated about the earth, the stars fixed to a giant sphere, the planets to another, spinning endlessly with the earth at their centers. When it came to the basic laws of motion, he was equally mistaken. He thought that an object, such as an arrow, must be continually pushed along to keep it moving and that the air did this by creating a vacuum in front of it. He knew nothing of relative motion and thought that if you stood near the back of a ship and dropped a ball, it would just fall straight down and hit the water as the ship moved out from under it. Anyone who has ever been on a ship knows that if you drop something heavy from the top of the mast, it will just land at the bottom of the mast, no matter how fast the ship is moving. Aristotle didn't think this was true, but perhaps he was never on a ship. Or if he was, he didn't think to test his grand theories with this simple experiment.

Neither did anyone else for a long time. It was almost two thousand years later, during the Middle Ages, that people like the French bishop Oresme began to question the teachings of Aristotle. Rather than rely solely on Aristotle, Oresme turned to nature itself to find the answers. This was a very radical idea back then and it didn't come any too soon, for in the meantime the writings of Aristotle had completely misled physics and astronomy and had left the door open wide for astrology, which is now known to be baseless.

By the time of Galileo, scientists began to understand the importance of doing experiments. An experiment is just a way of asking nature a question. When an answer to a series of questions is determined, it is possible to develop a model or a theory based on the data. This method is known as inductive reasoning. Once a model or theory is advanced, one can make deductions from the model to determine new facts, or suggest new experiments. A good theory will accurately predict the results of these new experiments. This is called deductive reasoning. These two opposite yet complementary approaches, inductive and deductive reasoning, formed the foundations of modern science. On them, others began to assemble knowledge and add it to the framework piece by piece. Science, heretofore shackled by the teachings of Aristotle, was able to progress, bit by painstaking bit, reaching its zenith with the publication of the *Principia* by Isaac Newton in 1686.

HEAD SHOT

As NASA convincingly demonstrated over three hundred years later, Newtonian physics is accurate enough to get to the moon and back.

Today, our planet abounds with satellite televisions and global communications; desktop computers more powerful than the world's best mainframe of a generation ago; and medicines that can cure polio, pneumonia, the Black Death, and leprosy. Our doctors treat cancer and AIDS with therapies undreamed of even ten years ago. Humankind can travel farther and faster and has access to more knowledge than any other time in history. There has been more progress and more innovations in the last century than in the five hundred millennia that preceded it. The greatest visionaries of the nineteenth century fell short in their attempts to imagine our world. Yet many of the inventions we depend on today are based on knowledge they had then.

This is because modern technology is based on science. The global communication satellites that provide television, news, and information would not stay in their orbits long unless the laws of physics specified precisely how high to place them and precisely how fast to fly them. Satellite orbits are based on the laws formulated by Isaac Newton. Newton, the greatest genius of his age, who stood "on the shoulders of giants" like Copernicus, Kepler, and Galileo, resolved the motions of the planets and the laws that govern them nearly four centuries ago. Men landed on the moon in large part because they knew exactly when, where, and how fast to go from the work of Isaac Newton.

But Newton is not the only giant relied on today. Many modern miracles depend on the work of mid-nineteenth-century physicist James Clerk Maxwell, who combined a series of disparate observations and formulas into four elegant laws describing electricity, magnetism, and light. These laws, improved and extended by the brilliant British mathematician Oliver Heaviside, ultimately led to the invention of alternating current, which allows the transport of electric power over large distances, and the creation of the radio. The ability to transmit signals through the air or through fiber-optic light cables or to perform delicate laser surgery stems from knowledge of electromagnetic theory, which originated over a century ago.

HOW SCIENCE ARRIVES AT THE TRUTH

As the second decade of the twenty-first century begins, modern civilization finds itself in the very heart of the information age. But information-age technology is possible today only because of work begun in the early 1900s by two men, Max Planck and Albert Einstein. In 1900, Planck found an irresistible explanation for the ultraviolet catastrophe, that is, the inability of classical theory to explain the radiation spectrum from a blackbody, an idealized object that absorbs all electromagnetic radiation falling on it. Blackbodies absorb and reemit radiation in a characteristic pattern called a spectrum. The spectral emission of a blackbody could not be calculated classically without getting unrealistic answers involving infinities. Planck found that a realistic value for the energy density of a blackbody cavity can only be obtained using the assumption that the wavelengths of light inside the cavity do not appear continuously but instead occur only in discrete intervals. This idea was so radical that Planck first presented it publicly as a kind of joke, but in private, he took it very seriously. Einstein's contribution was to observe that not only could light travel in waves, but it could also appear as discrete units of energy that he called photons. His photon picture successfully explained the photoelectric effect, the process by which a beam of light shining on a metal causes electrons to be ejected from its surface. Classical theory had failed utterly to explain this effect. These two iconoclastic propositions were the beginnings of quantum theory, and hence of modern physics. It is an understanding of quantum mechanics that makes possible the inner workings of the modern computer, the workhorse of the information age.

Computers today rely on devices made from materials called semiconductors, substances that sometimes conduct electricity and sometimes don't. Understanding how to switch them on and off is crucial to the operation of fundamental circuit elements like diodes and transistors. But this knowledge in turn depends on an understanding of the energy levels of electrons in solids, called band structure. Band structure can only be comprehended through the theory of quantum mechanics; without it, there is no way to know precisely how to make the basic components that store and process information in a computer.

HEAD SHOT

The quantum world is full of surprises. In classical theory, when an object, such as a ball, for instance, reaches a barrier, such as a wall, it must stop. But in the world of very small objects, like electrons, a whole new set of rules apply. An electron doesn't have to stop at a barrier; instead it can do something quite magical—it can go straight through the barrier and appear on the other side by a process called tunneling. Tunneling has no comparable analog in the classical world, the world of everyday experience. But in the bizarre and uncertain quantum world, where both the position and the momentum of an object can never be known precisely, tunneling is a very real and important phenomenon. It is so significant that it is central to the operation of the diode and the transistor.

The transistor, which was developed at Bell Laboratories in the 1930s, is a circuit element that lies at the heart of the modern computer. It allows a signal to be switched, thereby forming a bit, the basis for the binary calculation system used as the language of modern computers. Tunneling allows electrons to penetrate from the valence band to the conduction band in the semiconductors used to form a transistor. The phenomena is also crucial to the operation of the tunnel diode, an important device element with very fast switching times used in high-speed circuits and high-frequency oscillators, as well as to understanding the fundamental constraints in device design required to prevent current leakage in microelectronic devices.

As transistors and other circuit elements become smaller and smaller and computers become faster and faster, quantum phenomena play an increasingly larger role. The smaller a computer, the faster it can operate because although electron currents travel at the speed of light, this velocity is still finite; in the domain of very small devices and circuits called Very Large Scale Integration technology (VLSI), this transit time becomes significant. The information age exists today and progresses tomorrow because of key discoveries made a century ago.

Lasers were developed in the 1950s by Arthur L. Schawlow, then a Bell Labs researcher, and Charles H. Townes, a consultant for Bell Labs. The laser works by a process called stimulated emission of photons from atoms in quasi-stable atomic energy states present in a gas or a solid. But the theory of the laser was actually worked out in the 1920s by Albert Einstein. At that time, Einstein presented the theory of bosons, a class of particles that photons belong to, their statistics, as well as Einstein A and B coefficients, which describe how the process of stimulated emission proceeds according to the laws of quantum mechanics. Without this theoretical underpinning, the invention of the laser and the maser (the laser's microwave-beam counterpart) in 1958 would not have been possible. Men like Newton, Einstein, and Planck were able to so profoundly influence the modern world because they discovered and formulated truths about nature. They uncovered these truths using a method we call science.

Many have confidence in science today because it has filled the world with miracles. Any person from any age past would be stunned by our modern achievements and technology. But every single one of these advances has stemmed from an understanding of nature. Without it, humankind would be forever locked in the world of the Middle Ages, relying on the ancients for answers, doomed to eternal stagnation. Fortunately for modern society, science is a process that gets at the truth. If it were flawed, scientists of the past could have provided no assistance today. Though the sum of their knowledge was less, the pioneers of science uncovered fundamental truths that the modern world has validated and exploited.

The progress of science today revolves around a subtle interplay between experiment and theory. By accurately fitting consistent data to models, scientists are able to generalize these models to find the right rules, laws, and theorems that accurately describe the workings of the universe. An important and relevant example of how science works at its best is provided by one of the most famous scientific success stories of the millennium, the discovery of the law of universal gravitation.

Toward the end of the sixteenth century, Johannes Kepler, the greatest mathematician of his day, began to puzzle seriously about the correct orbits for the planets. He believed that the ancient Greeks, who had mastered geometry and trigonometry, provided the framework for the right model. The Greeks had been fascinated by the perfect solids and harmonies in nature, a view they described as "the harmony of the spheres." They believed the sphere to be "perfect," since all points on its surface were the same distance from the center. Kepler subscribed to the Greek view, writing that "the universe was stamped with the adornment of harmonic proportions," and he believed that the heavens could be understood in terms of simple numerical relationships.[11]

In Kepler's time, only six planets were known; Mercury, Venus, Earth, Mars, Jupiter, and Saturn. At the same time, only five regular solids—geometrical shapes with regular polygons as their sides—were known to exist. A regular polygon consists of a closed geometrical shape where all the sides are of equal length, like a square, or a pentagon with five equal length sides. Kepler began to imagine that the orbits of the six planets, perfect spheres centered on the sun, could be understood if the five regular solids were inscribed inside the six planetary spheres, supplying the necessary supporting structures. This would complete the mathematical program initiated by the ancients and lead to a mathematical model of the solar system.

Kepler began to attempt to construct his model out of paper. "The intense pleasure I have received from this discovery can never be told in words.... I shunned no calculation no matter how difficult. Days and nights I spent in mathematical labors, until I could see whether my hypothesis would agree with the orbits of Copernicus or whether my joy was to vanish into thin air."[12] But no matter how hard he tried, Kepler couldn't quite make the model work to his satisfaction.

But his efforts did not go unrewarded. The great Tycho Brahe, a Danish nobleman who had built his own astronomical observatory on the isle of Hven called Uraniborg, had learned of Kepler's work through his benefactor the Holy Roman emperor Rudolph II, and Brahe invited the gifted mathematician to work with him. Although Kepler at first had misgivings, in 1598 he packed up his family and journeyed to Prague to join the famous Brahe at his new laboratory at Benátky nad Jizerou.

Brahe had spent decades observing the heavens. He had stocked his laboratory with the finest collection of astronomical instruments ever assembled prior to the invention of the telescope. Brahe not only had the best instruments but also had painstakingly made the best naked-eye survey of the heavens ever conducted. His data was unrivaled. Kepler's goal was to analyze the exquisite data of Brahe, amassed over decades of careful naked-eye measurements. Kepler believed that together, he and Brahe could determine the correct model of the solar system.

But things did not work out as Kepler had planned. Brahe was reluctant to share his precious data with the precocious younger scientist. The two quarreled continuously over the eighteen-month period that they were together. However, after indulging heavily in wine at a banquet given by the baron of Rosenberg, and not looking after his health properly, Brahe soon found himself on his deathbed. On the last night of his life, Brahe bequeathed his marvelous heavenly observations to Kepler, repeating the words "let me not seem to have lived in vain" over and over again. Brahe need not have worried. Kepler would go on to immortalize him.

With Brahe's death, Kepler was appointed the new Imperial Mathematician, and with his new position he was able to persuade Brahe's reluctant family to release his data. Kepler persisted at the task of modeling Brahe's data well into the next century. Brahe had suggested to Kepler that he focus on the planet Mars, because this planet seemed hardest to reconcile with the model of circular orbits.

Kepler was convinced that the Copernican model was the right one and he was able to fit the data reasonably well to circular planetary orbits with the sun at the center, yet there were still small discrepancies. After three years of calculation, he thought he had discovered the correct parameters to characterize Mars's orbit. He was able to match ten observations to within two minutes of arc, about a thirtieth of a degree. This was quite precise, but two more of Brahe's measurements were still off by as much as eight minutes. Kepler was convinced of the accuracy of Brahe's data. It was the best set of naked-eye astronomical observations in history, unsurpassed until the invention of the telescope. In the end, Kepler found that he simply could not ignore the eight minutes of arc discrepancy and set about finding a new model.

Abandoning his belief in circular orbits clashed seriously with Kepler's religious beliefs. He was beginning to have grave doubts about the Divine Geometer. Eventually, in a flash of insight, Kepler tried an elongated circle, an ellipse, as the shape of planetary orbits, another one of the conic sections described by the ancient Greeks, and to his astonishment, it matched Brahe's astronomical observations perfectly. It was a short step from this success to the establishment of his three famous laws for planetary orbits, proposing that planets traveled in elliptical paths around the sun, swept out equal areas in equal times, and followed a precise relationship between the period of a planet's orbit and the radii of that orbit.

Galileo Galilei, the great Italian physicist who had invented his own telescope and turned it toward the heavens, did not embrace Kepler's work. While he was a firm believer in the Copernican model of the solar system, he could not bring himself to accept Kepler's radical idea of elliptical planetary orbits. Isaac Newton, however, could.

Newton was a student at Cambridge when, in 1665, the plague broke out in London. During that year, Newton was forced to remain at his home in the countryside. It was during this time that he began to study and master the mathematics of the ancient Greeks. He had especially studied the *Conics* of Apollonius of Perga, considered the most difficult of all ancient texts. From it, he made one of the most extraordinary breakthroughs in the history of mathematics; he invented the calculus.

The calculus is a branch of mathematics that deals with change in mathematical functions. That change can be strictly geometrical in nature or it can involve time as well. If it involves time, then it can be used to determine velocities and accelerations. The laws of motion precisely describe the change in position of an object as a function of time and can be expressed accurately and precisely in the mathematical language of calculus. When Newton plugged Kepler's elliptical orbits into his equations of motion for a planet revolving around the sun expressed using calculus, the correct law for the central force, in this case gravity, fell out. Using his novel mathematics he directly calculated that only an inverse square law for the operation of gravity could reconcile his equations of motion with the elliptical orbits of Kepler. Newton had not only discovered the correct law of universal gravitation, but he had mathematically and scientifically proved it as well. To many, this was a major turning point in scientific history.

This success story is a bright, shining example in the history of science because Kepler's painstaking work led directly to Isaac Newton's discovery of the correct law of universal gravitation. Kepler's single-minded pursuit of a perfect fit to the astronomical data not only led him to the correct solution but also ultimately to the establishment of the foundations of the science of physics. Had Kepler ignored the discrepancies between his model and the scientific data, or if he had refused to abandon his preconceived notions of what the planetary orbits should be, his name would likely have been lost to history. Instead, his impact on the history of science was greater than he could have imagined.

Ironically, it was the presence of a supernova in the skies of his era that allowed him to reach the right answer. While Kepler had at first tried nested perfect spheres for the orbits, because he believed, as did the Greeks, in the harmony of the spheres, the perfect and timeless interrelationship between mathematics, music, and nature, he later found that this view was flawed. The temporary appearance of a supernova in the skies of 1604, the last seen in the Milky Way galaxy, convinced Kepler that the heavens were not immutable and that the Greeks were wrong. Their view of an unchangeable, mathematically perfect universe could not have been right. The appearance of the supernova permitted him to abandon his preconceived notion of a perfect circle to represent orbits and move on to the ellipse.

Supernovas pose special challenges in science since they are transient events. They cannot be reproduced. They leave no tangible physical evidence behind, only lurking black holes from which not even light can escape. So scientists have come up with ways to deal with these tantalizing and significant occurrences that tempt and taunt us like Lewis Carroll's Cheshire cat. The supernova data turns out to be some of the most useful data in all of science. Why? Because Type Ia Supernovas are the standard candles of the universe.

If a lightbulb is examined up close, it appears so bright that it will hurt the eyes. But if the same lightbulb is observed from a distance, it appears faint. The intrinsic luminosity of the bulb remains unchanged; the difference is that the light emanating from it diffuses over distance. The same amount of light is passing through the surface area of the bulb. But as distance from that bulb increases, so does the total surface area that the light must pass through to reach the observer. Surface area increases as the square of the distance from the source. So, as that distance increases, the light appears to dim by that same ratio. So, a bulb observed at a distance of ten feet appears at a certain brightness, but if it is viewed from a distance of twenty feet, the same bulb will appear to be four times fainter. If observed from a distance of one hundred feet, that bulb will appear one hundred times fainter to the eye. The intrinsic brightness or luminosity of the bulb has not changed, but our perception of its brightness is altered based on our distance from the source.

The same phenomenon occurs on a vast scale in observations of the universe. If an object is observed radiating light, it is possible, in principle, to determine its distance provided its intrinsic luminosity is known. Therein lies the problem. The intrinsic luminosity of a star, for instance, is indeterminable, unless its distance is already known. This is a kind of Catch-22 in the universe. Type Ia supernovas, however, produce virtually the same radioactive output for each event. Moreover, these spectacular stellar explosions are extraordinarily bright, allowing for their observation in even very distant galaxies. This makes them the ideal "standard candles" of the universe. Since their intrinsic brightness is known, it is possible to calculate the distance of the galaxy where the supernova occurred based on the amount of light that reaches our telescopes on Earth. This, in turn, allows for the determination of the Hubble constant, a measure of how fast galaxies are receding from the earth based on their distance from it. Extrapolating backward, this constant can be used to determine the age of the universe.

HOW SCIENCE ARRIVES AT THE TRUTH

It is easy to see that the supernova data is crucial to the understanding of the cosmos. Supernovas, however, cannot be reproduced. A massive star explodes once in its lifetime and that's it. The event is observable for a period of about two weeks. After that, it's gone. In a distant galaxy, it leaves no remnant of itself behind, no telltale trace of its existence. Therefore, it doesn't fit into the standard method of scientific analysis. However, using the method of multiple independent instrumented records, scientists can nonetheless overcome this limitation and gain extremely high confidence in their observations of these crucial, yet fleeting, events.

If a supernova is observed in a distant galaxy, the team that first sees it will contact other astronomers around the world to see if they can also observe the event with their telescopes. If a bright object appears in the sky on two separate telescopes in the same place and at the same time, then the probability that either observation is in error is very small. Because for both events to be due to "random noise" or to, say, a "luminescent dust speck" on the telescope's mirror, it would require that both mirrors have the dust speck in *exactly the same place*. Since a telescopic mirror may be hundreds of square feet in area, the probability of this happening is astronomically small. If a third and fourth observatory see the same bright event in the same galaxy, this provides an even higher confidence in the result. And if other radiation detectors, like gamma-ray detectors, and radio telescopes see high energetic output from exactly the same position in the sky, then these observations taken together are considered even more powerful proof of a real astronomical event. If two instrumented methods searching in multiple spectral domains using different technologies can observe the same event, then this is considered very powerful evidence in science. It is exceedingly improbable that a pernicious random noise source could cause both independent instruments to "glitch up" so as to falsely "appear" as identical real signals in both records in exactly the same place and at exactly the same time.

Science depends on probabilities, not certainty. In fact, quantum mechanical uncertainty is woven into the very fabric of the universe.

HEAD SHOT

There is a finite probability that all the atoms in your body will fly apart and reassemble on the moon. The probability of that happening is very, very small, but it's not zero. The advent of quantum mechanics has forced scientists to view the world very differently than in the age of "classical" deterministic science over a century ago. Everything has a probability associated with it. As a result, if a scientific observation can achieve a very high probability of accuracy, as is possible through the use of multiple independent records taken at the time of the event, like astronomical film records or digital data recording, then it has scientific value and significance and can be reliably used to generate systematic models.

Even though a natural event is not reproducible, it doesn't mean it can't be studied scientifically. If science had to throw out the supernova data, for instance, all understanding of the universe, the origins of the big bang, the dark energy acceleration, and the fate of the cosmos would be lost. The field of cosmology would be back to square one. The use of multiple independent instrumented records to achieve scientifically valid data with high confidence is currently a successful and accepted way to make progress in modern scientific research.

Scientists study supernovas, earthquakes, and other fleeting events by employing the method of multiple independent instrumented recordings. The probability of two or more independent instrumented measurements erroneously showing the same result, at the same time consistent with a real phenomenon, is astronomically small. The use of multiple instruments to study short-lived phenomena allows scientists to have confidence in the validity of their results.

This then is the scientific program needed to determine the course of events that transpired in Dallas on November 22, 1963.

CHAPTER 5

THE MEDICAL EVIDENCE

You must remember, Humes and Boswell had never done medical-
legal autopsies in their careers. It [the autopsy] was really inept.
—Dr. Cyril Wecht, former head of the
American Academy of Forensic Sciences

Parkland Memorial Hospital was the major public hospital in Dallas in 1963 and remains so to the present day. Its emergency room was staffed by highly experienced trauma room physicians and support personnel. At 12:30 p.m., President Kennedy was rushed to Parkland at speeds approaching eighty miles per hour and arrived within minutes of being struck with multiple bullet wounds in Dealey Plaza. By 12:38 p.m., some of the most competent and skilled medical professionals in the country raced against time to try to save his life.

The scene at the entrance to Parkland Hospital became an unforgettably tragic and horrendous moment in the nation's history. The First Lady, Jacqueline Kennedy, covered in blood, cradled her husband's head in her arms, refusing to release him to the Dallas medical personnel. She rocked back and forth saying only, "They killed my husband. They killed Jack." Minutes passed before emergency personnel could persuade Mrs. Kennedy to release the president to their care.

Dr. Bill Midgett readied a gurney, in response to commotion and disruption inside the hospital, and brought it outside to the limousine. Officer H. B. McClain, who had ridden in the Dallas motorcade, walked to the car and attempted to get Mrs. Kennedy to release her mortally wounded husband. He succeeded. As Mrs. Kennedy rose in her seat,

Bill Midgett could see the president's head wound. He knew instantly that Kennedy's prospects for survival were hopeless. Nonetheless, President Kennedy was placed on the gurney and wheeled into the hospital with Mrs. Kennedy by his side.

Dr. Charles J. Carrico was the first doctor to examine Kennedy. Carrico noted that the president was ashen in color, showed no voluntary movements, but was still breathing, though irregularly. Entering the emergency room next was Dr. Marion Thomas Jenkins, followed quickly by Dr. Malcolm Perry and Dr. Charles Baxter. Jenkins recalled, "He had a death look. He was definitely on the way out."[1]

Two nurses removed the president's clothes and back brace, which had been secured with bandages. While a US senator, Kennedy had undergone major surgery on his back requiring two operations spaced four months apart.[2] Dr. Carrico felt underneath Kennedy with his hands, searching for wounds on the back of his body, but found nothing. Dr. Perry immediately noticed a small wound in Kennedy's windpipe. He knew from experience that in order to keep Kennedy breathing, he had to perform a tracheotomy over the wound. This was the only way to ensure that Kennedy could breathe properly without air leakage.

Dr. Malcolm Perry later described the throat wound as a small hole about three to five millimeters in diameter that appeared to be an entrance wound. Perry said he made a surgical incision laterally across the hole to support a tracheotomy but did not eradicate the original wound.

Dr. Carrico inserted a cuffed endotracheal breathing tube, attached to a respirator, into the small puncture wound in Kennedy's throat. Carrico positioned the cuff below the throat wound and inflated it. Although Kennedy's respirations improved slightly, they were still inadequate. His breathing sounds were diminished especially on the right side.[3] This indicated that Kennedy's respiratory integrity was compromised presumably by another hole or wound somewhere else in his respiratory system.

Three doctors, Charles Baxter, R. Paul Peters, and Robert N. McClelland, then inserted a tube into Kennedy's chest cavity to drain excess fluid that could have impeded breathing.[4] After the tracheotomy, Kennedy's pulse began to decline due to internal hemorrhaging. Dr. Perry began closed-chest massage while the remaining doctors debated performing open-chest massage of the president's heart. Kennedy's life was fading. Dr. Kemp Clark, a neurosurgeon, began to examine Kennedy's head wound. It was immediately clear to him that the wound was too extensive to treat.[5] Dr. Clark informed Dr. Perry, who was still performing chest massage, of the hopelessness of saving the president. In Dr. Clark's experience, no one with such a wound could survive. As Kennedy's pulse reached zero, the heroic efforts to resuscitate him were terminated. A priest was then summoned to administer last rites of the Catholic Church.

Despite the extraordinary and commendable efforts of the Dallas doctors, Kennedy was pronounced dead at 1:00 p.m. (CST). Nurses and residents wrapped Kennedy's body in white sheets and pillowcases together with a plastic cover.

At 1:33 p.m., Mac Kilduff, a Kennedy aide, announced the president's death to the world. By 1:40 p.m., Kennedy's body was placed in a bronze ceremonial casket for transport back to Washington, DC. By 3:10 p.m. CST, United Press International reported a quote from a Dallas emergency room physician: "Dr. Malcolm Perry, thirty-four, said 'there was an entrance wound below the Adam's apple.'"[6]

The Dallas doctors consistently reported a large blasted hole in the right rear portion of Kennedy's head. It is conceivable that with the bone in the side of his head shot away, unsupported skin may have flapped over Kennedy's ear creating the appearance of a wound extending back toward the rear of his head. The official hospital report indicated that "[t]here was a great laceration on the right side of the head (temporal and occipital), causing a great defect in the skull."[7] The temporal region of the skull is just forward of the ear, while the parietal bone lies just above the temporal bone, but not extending to the top of the skull. The occipital region lies behind the ear, at the rear of the skull.

A piece of Kennedy's skull was later found in the street in Dealey Plaza by a medical student at Texas Christian University, William Allen

Harper, while taking photographs at about 5:30 p.m. the day of the assassination. Harper took the fragment to his uncle, Dr. Jack Harper, who then took it to Dr. A. B. Cairns, chief of pathology at Methodist Hospital, Dallas, for analysis. Although this bone was originally identified as occipital (from the rear of Kennedy's skull), recent evaluation by Dr. Joseph Riley, a neuroanatomist, indicates that the bone is instead from the parietal (side) portion of the skull. This analysis is based on the presence and the pattern of vascular grooves on the inner surface of the bone fragment, indented impressions where blood vessels ran,

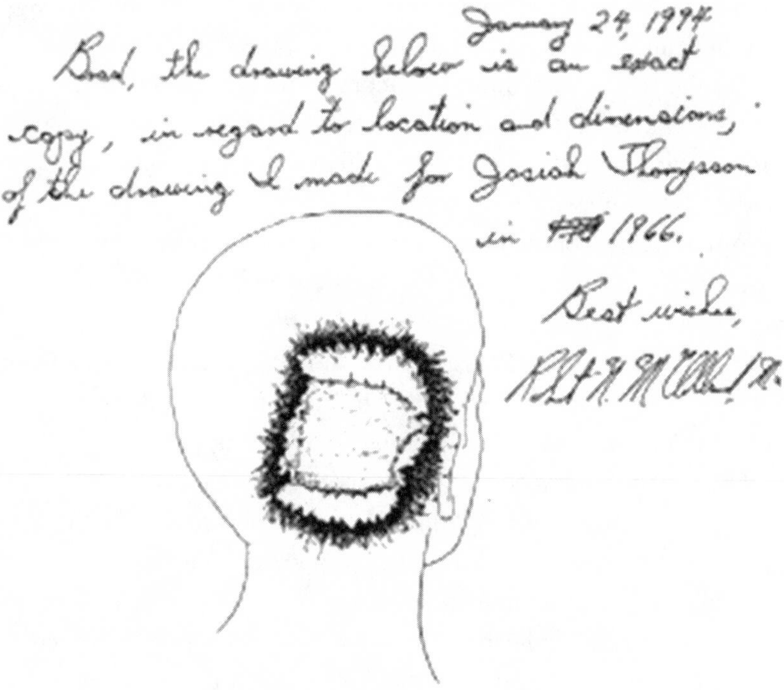

A pictorial representation of President Kennedy's head wound as described by Dr. Robert N. McClelland of Parkland Hospital.

Figure 6: Drawing by Dr. Robert McClelland of Kennedy's head wound at Parkland Hospital.

which is characteristic of parietal bone and not of occipital bone. Parietal bone has a typically smooth inner surface, gentle curvature, and relatively uniform thickness. In contrast, occipital bone is characterized by major variations on its internal surface, larger curvature, and consid-

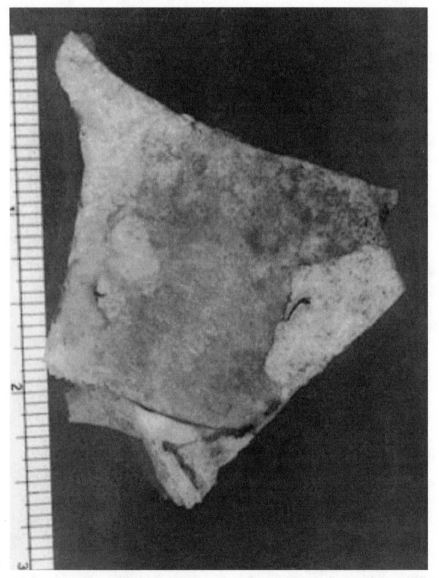

Figure 7: The Harper fragment, a trapezoidal piece of Kennedy's skull found in the street after the assassination.

erable variation in thickness. "Simply put, occipital bone doesn't look like the Harper fragment, but parietal bone does."[8] This result is consistent with the statement in the medical report from Dallas that the wound was located "on the right side of the head."

In violation of Texas law, Kennedy's body was removed from Parkland Hospital before an autopsy could be performed and transported to Air Force One. A heated argument broke out between the some of the Parkland medical personnel who wanted the autopsy performed there and Kennedy's people, who were determined to take the body back to Washington. Guns were drawn. Eventually, the Parkland staff backed down, and Kennedy's body was removed by an armed escort.[9]

However, Kennedy's body arrived at Bethesda Naval Hospital in a different casket, a gray military shipping casket. Kennedy's body was taken out of a black zippered body bag. An x-ray technician, Jerrol F. Custer, would later say that he had already made x-ray photographs of Kennedy's body and was returning from processing them when he saw Jackie Kennedy arrive with a throng of reporters and Secret Service agents. With her was the bronze ceremonial casket in an ambulance that was supposed to be carrying her husband's body.[10] This account was corroborated by another x-ray technician who was present that night, Lt. William B. Pitzer: "Kennedy's body was brought in through the back door in an unmarked ambulance. An official motorcade from the airport contained only an empty casket."[11]

The official autopsy began at 8:15 p.m. It took place in a small, crowded autopsy room filled with military and civilian personnel and federal agents. Admiral George G. Burkley, Kennedy's personal physician, was also present. Also in the room were two FBI agents, Francis X. O'Neill Jr. and James W. Sibert, who were in attendance to make a report of the autopsy. According to O'Neill and Sibert, preparation for the autopsy at the hospital had begun at 7:17 p.m., which indicated that Kennedy's body had arrived at the hospital fifty-three minutes earlier.[12]

Two of the three military physicians who were to perform the autopsy, Navy Cdr. James J. Humes and Navy Cdr. J. Thornton Boswell, were clinical pathologists with no firsthand experience with bullet wounds, and the third, Army Lt. Col. Pierre Finck, who did have some such experience, would later say that he had been hampered in his autopsy procedures by both military and civilian officials in the room.[13]

Assassination researcher Lamar Waldron, together with Thom Hartmann, has published two rigorously researched books on the Kennedy assassination. In his excellent work titled *Legacy of Secrecy*, Waldron reports that "the man really calling the shots [at Bethesda Naval Hospital] was Bobby Kennedy, from the family suite on the hospital's seventeenth floor."[14] Bobby was with a group that included Mrs. Kennedy as well as presidential aides Dave Powers and Kenneth O'Donnell. Since Bobby Kennedy had known and worked with both men for years, he surely would have trusted their firsthand accounts of a shooter from the grassy knoll.

As Waldron states, a number of people in the autopsy room reported that, working at Bobby Kennedy's direction, Admiral George Burkley "wielded a heavy hand at the autopsy on Bobby's behalf."[15] Agent Francis O'Neill later told investigators that there was "no question that Burkley was conveying the wishes of the Kennedy family." Jerrol Custer, an x-ray technician, reported that Burkley said, "I am JFK's personal physician. You will do what I say." A second laboratory technician, Paul O'Connor, went further, reporting that they performed only a "perfunctory examination" of Kennedy's internal organs because Burkley "kept yelling that the Kennedy family wanted just so much done, and that's all and nothing else."[16] Burkley was also giving orders to higher-ranking officers in attendance. Interestingly, as the only physician at the official autopsy who had also seen Kennedy's body at Parkland Hospital, Burkley would later state that he believed Kennedy had been killed by more than one gunman.[17]

As the autopsy began, FBI agents Sibert and O'Neill reported on events they saw there and what they heard the doctors say. The Sibert–O'Neill FBI report of the Bethesda autopsy states: "The president's body was removed from the casket...and placed on the autopsy table...it was also apparent that a tracheotomy had been performed, as well as surgery of the head area, namely, in the top of the skull."[18]

The autopsy itself focused on only two of Kennedy's wounds. The head wound was described as a gaping lesion in the right top portion of his skull. Humes examined this wound and found copious metallic bullet fragments on the right hemisphere of Kennedy's brain. This hemisphere appeared to have suffered an extraordinary degree of damage. No fragments were found on the left hemisphere of Kennedy's brain, which appeared to be intact.[19] This determination was confirmed by x-rays taken of Kennedy's skull. According to Humes, the bullet had shattered into about forty dustlike particles that appeared on the x-ray film like "stars at night."[20] In addition, a small 6 mm diameter bullet entrance wound was found in the rear of Kennedy's head. Humes believed that a high-velocity rifle bullet had entered through the rear of the skull, fragmented, and exited through the top of the skull, causing Kennedy's death.

Next, the wound in Kennedy's back was examined. The Dallas doctors had never seen this wound because they had never visually examined Kennedy's back. Dr. Humes's autopsy sheet diagram depicted the wound in Kennedy's back between his shoulder blades. The death certificate stated, "a second wound occurred in the posterior back at about the level of the third thoracic vertebra." Dr. Humes and other doctors probed the back wound with their fingers but could go no more than an inch. A metallic probe was then used, but no bullet could be located. The angle of the wound was found to be downward at an angle of 45 to 60 degrees. The location of the back wound was verified on the autopsy sheet by Burkley.[21]

Since no bullet and no path for a bullet could be found and the x-rays of Kennedy's body showed no bullet fragments inside the body, the autopsy doctors became puzzled. According to the Sibert–O'Neill report: "Inasmuch as no complete bullet of any size could be located in the brain area and likewise no bullet could be located in the back or any other area of the body as determined by total body x-rays and inspection revealing there was no point of exit, the individuals performing the autopsy were at a loss to explain why they could find no bullets."[22]

Agents Sibert and O'Neill contacted the FBI Laboratory and were told that an intact bullet had been found on a stretcher at Parkland Hospital. The agents informed Dr. Humes, who concluded that the external cardiac massage performed at Parkland caused the bullet to work its way back out of its entry point and fall onto the stretcher.[23]

However, Dr. David Osborne, a captain at the time but later promoted to admiral and the deputy surgeon general, told a congressional investigation that he saw "an intact bullet roll ... onto the autopsy table" when Kennedy was taken out of his casket. The bullet was later taken by the Secret Service.[24] This account was corroborated by Jerrol Custer, the x-ray technician who recalled that a "pretty good sized bullet" fell out of Kennedy's upper back.[25]

The doctors who performed the Bethesda autopsy never examined the throat wound because they thought it was only an oversized surgical incision for the breathing tube. It wasn't until the next day that Dr. Humes learned of the presence of the throat wound in Dallas. The large gash in Kennedy's throat, about 8 cm long, appeared to completely obliterate all traces of a bullet wound at the autopsy, in contradiction to the report of Dr. Perry. David Lifton, a former NASA engineer and early assassination researcher, thought that this was evidence of a prior probe of the wound in an attempt to locate or remove the bullet.

November 23, 1963's *New York Times* reported that "Dr. Malcolm Perry, an attending surgeon, and Dr. Kemp Clark, chief of the neurosurgery at Parkland Hospital gave more details. Mr. Kennedy was hit by a bullet wound in the throat, just below the Adam's apple, they said. This wound had the appearance of a bullet's entry."[26]

Perry had made an incision in this wound for the tracheotomy. However, that incision was at most 4 cm long, and so thin that the original wound was still visible. David Lifton reported asking Dr. Carrico about the incision Dr. Perry made. Carrico informed him that, "[p]robably—it would just be a guess—between two and three centimeters, which is close to an inch." When Lifton asked if the incision could have been as long as 4 cm, Carrico replied, "Oh, I really don't know, but that would probably be the upper limit."[27] Lifton then posed the same question to Dr. Charles Baxter, who responded, "Oh, it was roughly an inch and a half (3.8 cm)." When he asked Dr. Jenkins if the incision could have been three and a quarter inches long (8 cm), Jenkins replied, "No, I don't think so."[28]

But Humes's autopsy report read, "Situated in the lower anterior neck at approximately the level of the third and fourth tracheal rings is a 6.5 cm long transverse wound with widely gaping irregular edges."[29] This is not consistent with a scalpel incision, which would have been a thin slit. Moreover, Humes's testimony to the Warren Commission indicated that the wound was "7 or 8 cm in length."[30] The throat wound at Bethesda was both longer and wider than the incision in Dallas.

The Dallas doctors were experienced with gunshot wounds and were not likely to misidentify an exit wound as an entry wound. Exit wounds tend to be jagged because a bullet is usually tumbling upon exit. If Perry's original tracheotomy incision was later expanded laterally and vertically, it is a reasonable indication that a search for a bullet or bullet fragments lodged in the throat was conducted. Such a search would make sense if the wound was an entrance wound.

If a bullet or fragments were removed in this manner prior to the official autopsy, this would explain the absence of the appearance of bullet fragments inside Kennedy's body on the x-rays from the portable x-ray machine.

Even more problematic was the vast discrepancy in the descriptions of the head wound. The Dallas doctors and hospital personnel all placed the wound on the rear right side of his head. The autopsy doctors reported the injury on the top right portion of his skull. Humes had described it as a "longitudinal laceration of the right hemisphere which was parasagittal in position."[31] This referred to a line of tissue damage running parallel to the back-to-front centerline of the skull. At the end of the autopsy, as indicated in the Sibert–O'Neill report, the doctors concluded that one bullet had entered Kennedy's back and fell out during the process of cardiac massage, while another high-velocity bullet had entered the rear of Kennedy's head and fragmented prior to exiting the top of his skull.[32]

However, the Sibert–O'Neill FBI report clearly raised substantial questions as to the validity of the conclusions from the official autopsy.

Figure 8: Enlargement of Z333 taken moments after the fatal head shot. The inset shows Kennedy's autopsy photograph for comparison. Note the absence of damage to the rear and top of Kennedy's head after the final fatal shot. The two photographs appear noticeably different. Zapruder film © 1967 (renewed 1995). The Sixth Floor Museum at Dealey Plaza.

In particular, Sibert and O'Neill's report of comments made by the doctors at the official autopsy indicated their adeptness at recognizing the presence of surgery while simultaneously raising the specter of tampering prior to the autopsy: "[It] was also apparent that a tracheotomy had been performed, as well as surgery of the head area, namely, in the top of the skull."[33]

While the Dallas doctors had performed a tracheotomy in an attempt to save Kennedy's life, they did not perform surgery on Kennedy's head. Moreover, at no time did they ever describe a wound in the top of his skull.

An enlargement of the Zapruder film frame Z333, captured moments after the fatal head shot, shows no visible damage to either the rear (occipital) area, the top right rear, or to the top of Kennedy's head. The only visible damage is to the right side of his head, just above and forward of his right ear. Although a top view of his head would help

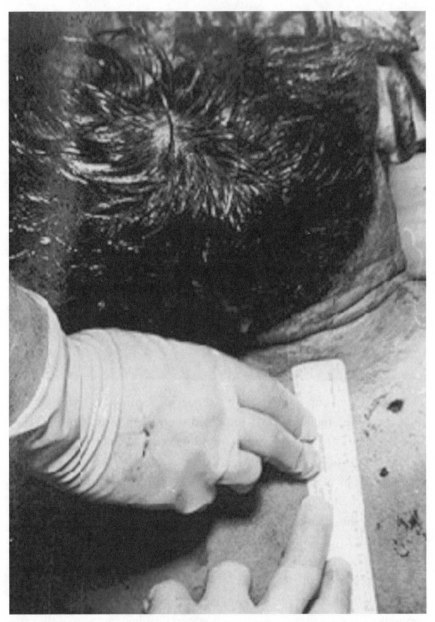

Figure 9: Autopsy photo of Kennedy's back wound. Note the absence of a wound in the neck. Note also the absence of a bullet wound in the back of the scalp. A flap of skin is apparent in front of Kennedy's right ear. Note also that no damage is apparent to the occipital region, in the lower rear of his head. Courtesy of Robert Groden.

resolve the issue definitively, the official autopsy report said that the extensive laceration on the top of his head was to the right of the centerline from back to front of Kennedy's head. Since the damage was to the right of centerline, and was substantial, some evidence would be expected to be visible on this film frame that clearly shows the right side of Kennedy's head from the top down. No visible damage of any kind is apparent at the top of his head, from the right ear to the top of the saggital crest.

Certainly, the top rear part of Kennedy's head, hair, and skull are intact, with no visible damage whatsoever to this region. Thus, there is a clear and apparent inconsistency with the Zapruder film, taken at the time of the assassination, and the description of the wounds to Kennedy's head provided at the official autopsy. The damaged area shown on the film is consistent, however, with parietal bone (from the side of Kennedy's head) found in the street after the assassination (the Harper fragment). This would be reasonable if the bullet struck Kennedy from the right front side and sheared off part of his skull on the side of his head just forward of his right ear.

One of the Dallas doctors, Dr. Kemp Clark, stated that Kennedy's head wound *could* be an exit wound, but that it was also possible that the wound was "tangential."[34] A tangential wound occurs when a bullet doesn't enter and exit the body at two separate locations, but instead strikes it from the side tangentially. In the case of a head shot, a bullet striking tangentially would produce a shearing force on the skull. Thus, a bullet striking from the front side could shear off the piece of parietal (side) bone, propelling it backward to the rear of the vehicle, and ultimately ending up in the street at Dealey Plaza.

Review figure 9. There is no evidence of an entry wound in the back of Kennedy's scalp. The back wounds appear to be in the shoulder area. The wound in line with the ruler is likely the wound that the doctors attempted to probe since it is clear that they are measuring its position in the photograph.

To make matters worse, the official x-rays of Kennedy's head appeared to show a large portion of his front right skull missing. In

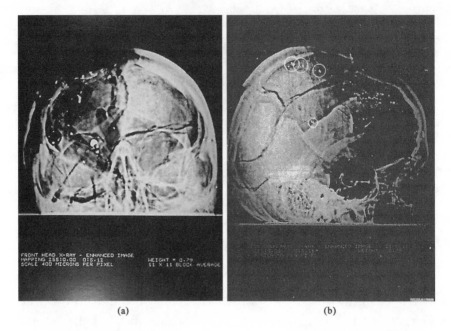

Figure 10 (a) JFK x-ray. Front view.   (b) Photograph of x-ray #2. Side view.

these x-rays, metallic bullet fragments are distributed in the front and the top portion of his cranium. However, no doctors at either Parkland or Bethesda described damage to Kennedy's face, right eye socket, or the front right portion of his skull.

The lack of any corroborative evidence for the skull damage in the official x-ray calls it into question. Because Kennedy's face was described as undamaged by witnesses, and this set of wounds is not described in either the autopsy reports or the reports of the Dallas doctors, we must conclude that these are not x-rays of Kennedy's head. This conclusion was first reached by Robert Groden, who was a photographic expert for the House Select Committee on Assassinations, and published in his book *The Killing of a President*.[35]

Unfortunately, there are enough discrepancies surrounding the official autopsy of Kennedy to call the entire set of facts into question. First and foremost, the data set is unstable. The head wound moves from the side of the head in the Zapruder film to the lower rear, occipital region, to the top rear of his head at the autopsy, to the top front of his head in the official x-ray. It's impossible to model data that changes as a function of time. The minimum requirement for any credible data set is that it be stable. Second, the reported data are inconsistent with other data. It's not consistent with what is observed on the Zapruder film and it's not consistent with the descriptions of the Dallas doctors. Last, there is a statement from a medical expert present at the autopsy suggesting that surgical alterations were performed between the time the body left Dallas and the time it arrived at the official autopsy in Bethesda, Maryland.

If a scientist were to attempt to formulate a model based on the data set from the official autopsy, say, of the head wound, what exactly would he model? Should he model the wound in the side of the head described by the Dallas doctors, the wound in the rear as shown on the Dallas doctors' drawing, the wound in the right rear and top of the head that was described in the Bethesda autopsy, or the wound in the top front of the skull as seen on the official x-ray? Real data can't change as a function of time. It can't change as Kennedy's body moves from one place to another. No conceivable model could be formulated for head wounds sustained at the assassination based on the medical evidence because the inflicted wound kept changing location on his head. Where did it start and where did it end?

One cannot say with absolute certainty whether the problems with the medical data are due to human error or to deliberate sabotage, although it is likely that both occurred in this case. From a scientific point of view, it doesn't matter. All one knows is that the data is unstable and inconsistent and therefore must be reviewed. If an experiment exhibits these features, it has to be reproduced. If physical evidence falls into this category, it has to be reexamined exactly as the original Piltdown skull was reevaluated when it became clear that it was inconsistent with other data points, newly found fossils that exhibited opposite traits.

Unfortunately, in the case of Kennedy, it is not possible to reexamine this original data. The brain is no longer extant. Its present location is unknown.[36] The soft tissue is no longer extant. A reexamination of the skull and the bones would require exhumation of Kennedy's body, something that is not likely to ever happen. In a case where data are suspect and can't be reviewed, they must be discounted.

Sadly, the medical evidence gleaned from the autopsy is a substantial data set in an important case, and it is tempting to attempt to somehow salvage something from it. It is potentially conceivable that some data from the Bethesda autopsy may be original wounds inflicted at the time of the assassination and subsequently unaltered. However, formulating models based on questionable data is a perilous endeavor. Any model based on this data is just as likely to be misleading as it is to be helpful. However, if data at the official autopsy can be corroborated by other reliable data, it's possible to consider that some findings may have been real, that is, unaltered wounds inflicted at the time of the assassination.

The wound in Kennedy's back appears to have some corroboration from other sources. There were corresponding holes in both Kennedy's jacket and shirt that matched the position of this wound on his body. Secret Service agent Glen Bennett was riding in the car behind Kennedy. He claimed to have seen a shot hit Kennedy about "four inches down the right shoulder."[37] Agent Roy Kellerman, seated in the limousine in front of Kennedy, heard Kennedy exclaim, "My God, I am hit," after the first shot.[38] This would have had to have been a wound other than the throat wound, because it is highly unlikely that Kennedy could have spoken with a bullet wound in his trachea near his larynx. When Dr. Carrico inserted a tracheotomy tube into the wound in Kennedy's throat, it wasn't holding air. There was a leak somewhere else in Kennedy's system. This could have been caused by a back wound that penetrated the president's lung or trachea as indicated in the official autopsy report: "This missile [that caused the back wound] produced contusion of the right apical parietal pleura of the right upper lobe of the lung."[39] This is consistent with the problems the Dallas doctors were having in maintaining integrity of Kennedy's respiratory system. This additional corroborative evidence suggests that the back wound was likely inflicted at the time of the assassination.

The small wound reported by the autopsy doctors in the back of

Kennedy's head is less certain. No substantial corroborative evidence for this wound exists. Although there appears to be a small forward motion of Kennedy's head between frame Z312 and Z313 of the Zapruder film, this motion could have been caused by deceleration of his limousine. Many witnesses on the scene reported that the car slowed down, and it is likely this was the case, as Secret Service agent Clint Hill was able to catch up with it. Some photos taken at the time show the limousine's brake lights on. Further, neither the Dallas doctors

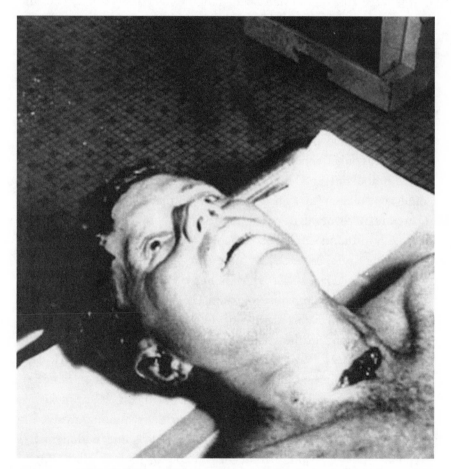

Figure 11: Death stare of Kennedy taken at Bethesda. Note that no wounds to his face are present and that the skull above his right eye is intact. Note the presence of the throat wound just below the Adam's apple.
Courtesy of Robert Groden.

nor any of the medical personnel, including the ones who wrapped the body and prepared it for transport, reported seeing this wound at Parkland Hospital.[40]

If this wound were inflicted at the assassination it could not have been from the Mannlicher-Carcano, a military rifle. The weapon fired hardened military jacketed rounds as required by the Geneva Convention. A 6.5 mm round fired from behind Kennedy would have penetrated his skull and exited through the front of his face or eye socket.[41] If this large rifle round had been frangible or hollow point, it would have exploded on impact, effectively blowing Kennedy's head off. This was proven by live-fire tests conducted by the Discovery Channel in 2008, where a .30-caliber Winchester fired a frangible round at an anthropomorphic dummy representing Kennedy's head. This frangible rifle round essentially obliterated the dummy's skull.[42]

The final problem with the conclusion that Oswald inflicted the small head wound from behind is that this wound, as measured at the autopsy, was only 6 mm in diameter. This was too small to be inflicted by the 6.5 mm diameter Mannlicher-Carcano round. A wound to the skull could not collapse back in to make the entrance hole appear smaller. Therefore, this wound, if it were real, was likely inflicted by a small-caliber 5.6 mm round fired by a weapon like the AR-15, a high-velocity, small-caliber rifle. An excellent discussion of this wound is provided in the book *Mortal Error* by Bonar Menninger, which describes the conclusions of gunsmith and ballistics expert Howard Donahue. After careful analysis of the available data, Donahue believed that the bullet that caused the wound in the back of Kennedy's head was produced by a small-caliber (.22) round from a high-velocity rifle like an AR-15.[43] Although I find the central thesis of this book (namely, that a Secret Service agent accidentally inflicted this wound with his AR-15 from the limousine trailing Kennedy) to be untenable, the reasoning involved in Donahue's ballistic analysis of Kennedy's head wound is nonetheless excellent. An AR-15 firing from the window or roof of a building in Dealey Plaza could have produced the small fragments observed in Kennedy's brain without exiting through his forehead or eye socket.

THE MEDICAL EVIDENCE

Could this wound have been inflicted after Kennedy's body left Dallas? If other alterations were performed on Kennedy's wounds in secret, then this could have been artificial as well. If some alterations to the president's body were performed, then it is impossible to know the full extent of the changes. A .22-caliber handgun with a silencer could have inflicted the rear head wound after the fact. But couldn't the doctors at the official autopsy have determined if a bullet wound was caused after death? The answer would be yes, provided they had sufficient time and opportunity to conduct the investigation as needed to determine the parameters of the gunshot wounds. According to Dr. Pierre Finck, the only physician in the room who had expertise related to gunshot wounds, neither opportunity was afforded them the night of the autopsy.

Could there have been bullet fragments in Kennedy's head? The official x-ray photo released to the public is clearly not that of Kennedy's head. However, technicians at the autopsy stated that bullet fragments were found on the x-rays they performed there. Other doctors found fragments when probing the wound. It would be extremely difficult to remove copious bullet fragments from someone's brain tissue if they had been there from the assassination, even if someone had time and opportunity prior to the official autopsy. Similarly, the introduction of numerous metallic fragments into tissue artificially would have been extremely difficult to accomplish. In principle, a shotgun blast would do the trick, but it would not only be very noisy and noticeable, it would do extensive damage to Kennedy's head, even more than that observed at the autopsy. Therefore, either introducing or removing abundant metallic fragments in a cadaver's brain would have been a major and virtually impossible task to complete given the time constraints confronting someone altering Kennedy's wounds prior to the official autopsy at Bethesda. Therefore, we might reasonably conclude that bullet fragments may have been in Kennedy's brain, although we cannot say anything with certainty concerning the precise distribution of those fragments in his head, other than to say the right hemisphere appeared damaged and contained copious fragments, while the left hemisphere of the brain appeared undamaged.

The most that can be said is that some of the wounds described at the official autopsy may have been real, inflicted in Dealey Plaza. Because of the overwhelming evidence of falsification, we can, however, conclude nothing with certainty based on evidence from the official autopsy.

The case of David Baltimore illustrates an important point about altered data. In 1986, a team of scientists, led by Baltimore, published an article in the journal *Cell* reporting new results associated with the functioning of the immune system. However, the inability of other researchers to reproduce some of the reported results led to an accusation of falsifying data in a published journal article. Tampering with or altering data is anathema in science. This is the one thing considered unforgivable by the scientific community because raw data are the stock and trade of the scientist. Many journal articles are read, digested, and modeled by other researchers in the field. Key findings are often cited in future publications, and generalized scientific hypotheses, models, and theories are based on them. False data lead researchers down the wrong path, waste valuable time and resources, and ruin careers. The harm done by even a single falsified result can be inestimable.

The Baltimore case dragged on for years. Baltimore himself, a distinguished scientist who was appointed president of the California Institute of Technology (Caltech) in 1997 and who was awarded the National Medal of Science in 1999 for his numerous contributions to the science, was never accused of wrongdoing. A member of the team who performed the research was under suspicion. However, the investigation centering on the falsified data bore Baltimore's name, an unfair and unfortunate consequence. It highlights the salient point that it doesn't necessarily take a massive conspiracy to alter or fabricate data; one person can do the job just fine. Moreover, the culprit can often do it without even coworkers and corroborators being aware of the alterations. This is what makes data distortion so pernicious.

In his book *Best Evidence*, engineer and assassination researcher David Lifton has published a theory that Kennedy's body had been altered prior to its arrival at Bethesda Naval Hospital.[44] This theory has been widely criticized and ridiculed on the basis that no reasonable opportunity existed for the body to be surgically altered. The body was monitored at all times during transport and when it arrived at Bethesda at 7:17 p.m. on November 22, 1963, it was attended by two FBI agents, Sibert and O'Neill. Therefore, it doesn't seem possible that Kennedy's body could have been stolen and altered prior to the official autopsy at Bethesda.

However, Lifton was convinced he had found a plausible chain of events that could account for his theory regarding the alteration of Kennedy's body prior to the official autopsy. He believed that Kennedy's body was removed from its casket aboard Air Force One, during the period in which Johnson was sworn in as president. Mrs. Kennedy was at the swearing-in ceremony at Johnson's behest, and therefore not attending Kennedy's body at the time. The entire time window may have been about fifteen minutes, enough time for Kennedy's body to be removed and relocated aboard the plane. When Air Force One arrived at Andrews Air Force Base in Maryland, the body was secretly removed through a rear door and placed on a helicopter, where it was taken to Walter Reed Army Hospital. There the body was surgically altered and taken to the Bethesda Naval Hospital by ground transportation. This explanation is consistent with the arrival of Kennedy's body at the Bethesda Naval Hospital in a casket different than the one in which he was originally placed in Dallas. Enough time is available in this scenario between the approximately 6:00 p.m. arrival of Air Force One at Andrews Air Force Base and the arrival of Kennedy's body at Bethesda at about 6:53 p.m., provided a helicopter was used in Kennedy's transport. However improbable this explanation may seem, the fact remains that this sequence is within the realm of possibility.

There is a class of people who make their living on the basis that no one can figure out how they do what they do. David Blaine is a master illusionist. He has a television special called *Street Magic* in which he levitates, teleports objects, and reads minds in full view of strangers he meets on the street for the first time. David Blaine does not violate the laws of physics. But he is one incredible magician. His success hinges on the reality that almost no one can figure out how he does his tricks. Magic is a very closely guarded secret, and the key element of a great illusion can be an extremely valuable commodity.

There are other magicians who can make elephants and cars disappear. David Copperfield has a trick where he appears to teleport himself through the Great Wall of China. He has another great illusion during which he makes a 747 Jumbo Jet disappear as it rests on the tarmac while a live television audience observes it from all sides. So, if one asks if it were possible to pull sleight-of-hand or use misdirection to make Kennedy's body disappear, sneak it off the plane, alter it, and return it, the answer would have to be in the affirmative. Whether Lifton's scenario is the correct one, or if Kennedy's body were surreptitiously altered in a surgery room at Bethesda before the official autopsy as a national security "pre-autopsy," or if it was done in the ambulance or helicopter on the way there isn't ultimately important. The salient point is that time and opportunity were available for the alterations to be accomplished prior to the official autopsy. If the impossible is ruled out, then what remains must be the truth.

The staff acting to redirect Kennedy's body may reasonably have done so believing they were acting in the nation's best interest. It was important to provide a decoy for Kennedy in the event that a person or group, national or international in origin, may have attempted to intercept his body. Everyone participating in this type of misdirection could believe he was acting in a way completely consistent with national security. Even the act of taking the body to Bethesda Naval Hospital via Walter Reed would not have been considered to be in any way suspicious by the people involved in the transport. Clandestine activity is part of their job description when national security interests are at stake. Even with a complex scenario like Lifton's, all that was needed was one person to actually perform the pre-autopsy surgery or alterations. The number of people involved in and cognizant of such an alteration plan could therefore have been surprisingly small, and one of the key reasons behind the surrounding secrecy that has been maintained for over forty years.

THE MEDICAL EVIDENCE

It is overwhelmingly clear that some type of tampering occurred prior to the official autopsy conducted at 8:15 p.m. A portion of the top of the skull was removed. The wound in Kennedy's throat was enlarged and expanded beyond the tracheotomy surgery performed in Dallas. From a scientific standpoint, it doesn't matter whether it is possible to find a clear opportunity for alteration that everyone can agree on. All that is known is that over six hours elapsed between the time Kennedy's body was loaded onto the plane in Dallas and the official autopsy was performed in Bethesda. Something happened to Kennedy's body during that period. It may not be possible to determine exactly what happened or when, but it is known that the appearance of Kennedy's head and throat at the official autopsy were dramatically different than they appeared in Parkland Hospital and on the Zapruder film. The statement ringing across the decades from the Sibert–O'Neill report, that there was "surgery in the head, particularly in the top," is damning.[45] Lawyers and politicians might ignore that statement. A scientist never would.

The entire official autopsy and everything subsequent to it must be called into question. It's impossible to know the full extent of the alteration to Kennedy's body. Scientifically, the evidence is corrupted and therefore must be discarded. Building a model on corrupted data is a sure way to get the wrong answer. In the final analysis, Kennedy's corpse is America's Piltdown Man.

CHAPTER 6

HOUSE SELECT COMMITTEE ON ASSASSINATIONS
THE ACOUSTIC EVIDENCE

Someone apparently does not want us to see the evil, hear the evil, and certainly does not want us to talk about it.... After all, if a President is eliminated ... isn't the whole form of our government in direct danger if we cover up the political motivations and go on as if nothing has happened?

—Rep. Thomas Downing of Virginia (1975)

Dissatisfaction with the Warren Commission report ultimately led to a reopening of the investigation by the House Select Committee on Assassinations (HSCA) in 1976. By this time, appearances of the Zapruder film on national television in 1975 had produced a groundswell of public support to revive the inquiry. The House Select Committee was originally established to investigate not only JFK's death but also the assassination of civil rights leader Dr. Martin Luther King Jr.

The HSCA began inauspiciously. After a year of infighting, a new chief counsel was selected. G. Robert Blakey, a professor of law and director of the Cornell Institute on Organized Crime, wasted no time taking control of the investigation. He chose witnesses, decided which leads to follow, picked the expert panels, hired and fired staff members, and set the committee's agenda. Because by this time many witnesses to the assassination had either died or experienced difficulty recalling events from fifteen years earlier, the committee decided to rely on scientific evidence if at all possible. It was believed that this approach was more likely to arrive at the truth than other methods under consideration.

HEAD SHOT

The committee performed extensive reevaluations of the medical and the physical evidence associated with the assassination. They looked carefully at the ballistic evidence, believing that neutron activation analysis had indicated that the two bullet fragments found in Kennedy's limousine were matched to Oswald's Mannlicher-Carcano rifle. Because the committee believed that the bullet found on the stretcher at Parkland Hospital had definitely wounded Connally, at least one of the fragments found in the front seat of the limousine must have struck Kennedy in the head. The HSCA concluded that the fatal head shot struck Kennedy from behind, entering the upper rear portion of his head.[1] Figure 12 illustrates their interpretation.

Dr. James Humes, the head of the Kennedy autopsy team, was questioned by the committee in its series of public hearings that lasted from late July 1978 to late September 1978. Many scientific and technical experts had been called to testify in the televised investigation. Although the hearings were extensive, Dr. Humes was never asked why he had waited fifteen years to determine that a bullet had entered in the upper rear of Kennedy's head, four inches above where the other two doctors on the team had placed it.[2] The committee also did not ask for details concerning the military authorities who were present at the autopsy and who had purportedly directed its substandard work. Many of the assassination researchers watching the televised hearings believed that the testimony was "orchestrated" and that it followed "carefully selected lines of investigation."[3]

Ultimately, the committee located an old radio recording, called a Dictabelt, taken from an open mike in Dealey Plaza. The existence of the Dictabelt recording was brought to the attention of the committee by longtime assassination researcher Mary Farrell, who supplied a taped copy of the recording to the committee.[4]

The original Dictabelt recording was made on November 22, 1963. Also, a tape recording of both channel 1 (Dictabelt) and channel 2 (Audiograph) were made. The tapes and original recordings were given to a committee investigator in March 1978 by Paul McCaghren, who in 1963 was a Dallas police lieutenant. Officer McCaghren had submitted his reports on these recordings to Chief Jesse Curry. In 1969, a newly appointed chief of police had ordered that a locked cabinet outside his office containing reports and materials concerning the assassination be opened. Among the items were the Dictabelt recordings and tapes of the November 22, 1963, dispatch transmissions. McCaghren, who in 1969 was director of the Intelligence Division of the Dallas Police

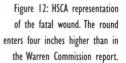

Figure 12: HSCA representation of the fatal wound. The round enters four inches higher than in the Warren Commission report.

Department, had then taken custody of the materials and kept them until releasing them to the committee's investigator in 1978. No evidence was found to indicate that any of the materials had been tampered with prior to their acquisition by the committee.

Channel 2 recordings apparently had not been in use at the time of the assassination. However, the recordings of channel 1 had been made by a motorcycle microphone that had been "stuck open" while in Kennedy's motorcade during the assassination. The HSCA now sought to determine if noises on the tape could be consistent with the sounds of gunfire in Dealey Plaza.

The HSCA commissioned the acoustics firm of Bolt, Beranek & Newman, Inc. (BBN) to perform a scientific analysis on the Dictabelt recording. This firm had previously successfully utilized acoustical analysis to determine the events that transpired during the Kent State shooting incident in 1970. Their acoustical analysis was later used as evidence presented to a grand jury to determine which national guardsman had fired first. BBN was also appointed by Judge John J. Sirica to serve on a panel of technical experts to analyze President Richard Nixon's Watergate tapes.

Led by their chief scientist, Dr. James Barger, BBN converted the sounds on the tapes to digitized waveforms. They then ran the waveforms through electronic filters to eliminate repetitive background noise like the sounds of the motorcycle piston firing. The firm then examined the processed waveforms for "sequences of impulses." Their analysis indicated that there were six sequences of interest, spaced together within an eleven-second period recorded on channel 1, which could be consistent with the sounds of gunshots.[5]

The BBN team subjected these sequences to further analysis. The chief scientist, Dr. James Barger, after careful examination of the channel 1 tape, concluded that at least six loud noises on the tape were consistent with gunshots. He recommended to the committee that further tests be conducted, in particular live-fire tests in Dealey Plaza.

Blakey surprised committee members with the information that the sounds were consistent with at least one shot from a second location, the grassy knoll, and that one shot came 1.66 seconds behind another, an impossibility for Oswald since FBI tests had demonstrated that at least 2.3 seconds were required between shots with the Mannlicher-Carcano. Barger had recommended that based on the BBN analysis, live-fire testing in Dealey Plaza would be required to determine the source and the direction of the shots that appeared on the tape.

Due to the significance of these findings, it was deemed appropriate to subject the BBN analysis to an independent review prior to authorizing the live-fire testing.[6] It was now clear to the committee that if the BBN analysis were correct, they had scientific evidence for multiple shooters in Dealey Plaza, and therefore substantiation for a conspiracy to kill the president. As such, it was imperative to ask independent experts to corroborate the conclusions of Dr. Barger before proceeding to live-fire testing.

The committee therefore contacted the Acoustical Society of America for recommendations of people qualified to review the BBN analysis and the proposed Dallas live-fire comparison test. The society recommended a list of individuals, and the committee ultimately selected Professor Mark Weiss of Queens College of the City University of New York and his research associate, Ernest Aschkenasy. Professor Weiss had worked on a variety of acoustical projects. He had served as one of the technical experts appointed by Judge John J. Sirica to examine the White House tape recordings in conjunction with the Watergate grand jury investigation. Aschkenasy had specialized in developing computer programs for analyzing large volumes of acoustical data.[7]

Weiss and Aschkenasy reviewed Barger's analysis and conclusions. They found that Barger's analysis was valid and his conclusions supported by the evidence on the tape. They concurred with his recommendation to conduct live-fire tests in Dealey Plaza to determine the origin and direction of the gunshots and they approved his plan for acoustical reconstruction.[8] After the independent corroboration by two more experts, live-fire testing was authorized.

On August 20, 1978, three Dallas police sharpshooters fired a total of fifty-six live bullets into three piles of sandbags located along the motorcade route on Elm Street. A Mannlicher-Carcano rifle was used to fire shots from both the Texas School Book Depository as well as from the grassy knoll. Since the weapon supposedly fired from the grassy knoll was unknown, a pistol was also fired from this location. Four target locations were chosen based on the position of Kennedy's limousine along the motorcade route as well as the location of the shot that struck a bystander, ricocheting from the street and hitting James Tague. An array of thirty-six microphones positioned along the route eighteen feet apart were used to record gunshots from the sixth floor of the book depository and from the fence along the grassy knoll.

During the live-fire tests, shots were fired without the telescopic sights, using only the iron sightings on the rifle. Tests performed by the Warren Commission showed that the fastest possible time between shots was 2.3 seconds. However, the tests for the Warren Commission were based on the assumption that Oswald had used the telescopic sights found on his rifle. The HSCA study found that two shots could be fired within 1.66 seconds provided that the telescopic sights were not used.[9] However, for this value to have represented Oswald's time between shots, it would have required him to have taken the time to reattach the scope after firing the shots and before hiding the rifle behind boxes. This would have been a crucial time for an assassin to make his escape. Moreover, Oswald was spotted and confronted only ninety seconds after the assassination in the depository lunch room by Officer Marion Baker. This would have been a very difficult for Oswald to accomplish if he had to take an additional twenty to thirty seconds to reattach a telescopic sight to his weapon. It is similarly difficult to believe that an assassin would not have maximized his chances of success by using the telescopic sight, in an effort to achieve accuracy, rather than rely on an iron sight in an effort to set a world-speed record between shots. Since only one shot must hit to be effective, accuracy is far more important than speed. The testimony of eyewitness Arnold Rowland that he had seen a man standing back from the window with "a fairly high-powered rifle because of the scope" supports the contention that the assassin in the depository used a telescopic sight. Lastly, this low time value between shots conflicts with statements made by the marksmen who performed the Warren Commission tests. In particular, they stated that 2.3 seconds was "the fastest the bolt could be operated." Although some marksmen are adept at "sighting around the scope," essentially pointing the rifle with the scope attached, the speculation that the shooter in the depository could have operated the bolt of the Mannlicher-Carcano and fired accurately faster than 2.3 seconds between shots is questionable.

The principle behind the acoustic reconstruction was based on the timing of shock waves and their reflections off buildings and structures in Dealey Plaza. When a projectile exceeds the speed of sound, as in a rifle shot, it produces a shock wave, a pressure wave, commonly known as a sonic boom. An example is the cracking of a whip; the tip briefly exceeds the speed of sound to produce the characteristic snap. The pattern of shock waves with time is called an acoustical signature, which differs from one sound source to another due to echoes and reflections off structures in the area. These echoes create characteristic patterns that can be used to determine the origin of the shock waves.

When someone yells loudly in an enclosed area, say, in a canyon, echoes can be heard reverberating off the walls of the canyon. Sound waves are produced by pressure spikes, or small shock waves, racing through the air. When those shock waves encounter a barrier, such as a rock wall or a concrete building, part of that shock wave bounces off the wall and can be heard as an echo. Sometimes multiple bounces will occur and the echo will appear to repeat and die off. The later echoes come from solid structures that are at a greater distance from the listener.

In Dealey Plaza, the sounds of gunshots would produce similar echoes. When recorded and captured on a specialized electronic device like an oscilloscope that converts sound patterns into pictures, these echoes appear as "acoustical waveforms" and appear as unique signatures of sound-producing events. In the case of a rifle shot in Dealey Plaza, the acoustical signatures would differ based on the origin, direction, and velocity of the shot, as well as the location of the recording microphone. The echo patterns would depend on the timing of sound reflections off of buildings and other structures and obstructions in the plaza. Sound travels at about 1,125 feet per second in air at sea level at 68° F. Given the distances in the plaza, these "echo" reflections would arrive at the microphone, shifted by milliseconds (thousandths of a second) determined by the exact location of the sound origin and receiving microphone, as well as the air temperature in Dealey Plaza. The precise velocity of sound varies with temperature. Because of their uniqueness, the analyzed sound patterns would provide an "acoustical fingerprint" of noise-generating events in Dealey Plaza.

For a microphone attached to a moving vehicle, like a motorcycle, a complication arises. That microphone would be moving along with the motorcade and therefore changing its location in Dealey Plaza during the time of the shots. Therefore, the precise location of the recording micro-phones with respect to the exact timing of the shots could not be known prior to setting up the live-fire test in Dealey Plaza. For this reason, an array of microphones had to be positioned along the motorcade route on Elm Street. Ideally, it would have been best to position the microphones as close together as possible. A one-foot spacing would have required 648 microphones, adding impractical expense and complications to the live test as well as to the amount of data that would have to be analyzed.

A recording was made of the sounds received at each microphone during each test shot, making a total of 432 recordings of impulse sequences (36 microphone locations times 12 shots) for various target-shooter-microphone combinations. Each recorded impulse sequence was then compared with each of the six impulse patterns on the channel 1 Dictabelt recording to see the degree to which significant points in each impulse pattern matched. The process required a total of 2,592 comparisons (432 recordings of impulse sequences times six impulse patterns).[10]

The time of the arrival of the impulses, or echoes, in each sequence of impulses was the characteristic being compared, not the shape, amplitude, or any other characteristic of the impulses or sequences. If a point (representing time of arrival of an echo) in a sequence of the 1963 dispatch tape could be correlated within 6 milliseconds (ms), or six-thousandths of a second, to a point in a sequence of the recon-struction, it was considered a match.

A 6 ms "window" was chosen because the exact location of the motorcycle was not known. Since the microphones were placed 18 feet apart in the 1978 reconstruction, no microphone was expected to be in the exact location of the motorcycle microphone during the assassina-tion in 1963. Since the location was unlikely to be exactly the same, and the time of arrival of the echo is unique at each spot, the 6 ms "window" would allow for the possibility that the motorcycle was near, but not precisely at, one of the microphone locations used for the reconstruc-tion. Those sequences of impulses that had a sufficiently high number of points that matched (a "score" or correlation coefficient of .6 or higher) were considered significant. This allowed the team to select out matches that were above the random noise background.

Correlation is a mathematical function that determines the degree to which two variables are related, or track each other. For instance, the success of a baseball team is inversely related to the earned run average (ERA) of its pitching staff. The lower the ERA of the staff, the more wins the team is likely to have. Even though the two numbers have an inverse relationship, they are nonetheless strongly correlated, as any baseball fan knows. A correlation factor of 1.0 would mean a perfect match between the waveform on the tape and the impulse sequence found in the Dealey Plaza live-fire test. A correlation factor of zero would mean that the two results were totally unrelated.

When the BBN team performed their analysis of the acoustical waveforms, they found something extraordinary. When they compared the impulse sequences from the acoustic reconstruction to the sequences on the original Dictabelt recording, they found a number of significant matches. When the locations of the microphones that recorded matches in the 1978 reconstruction were plotted on a graph of time versus distance, it was found that the location of the microphones that recorded matches were clustered around a line on the graph that was consistent with the known speed of the motorcade (11 mph), as estimated from the Zapruder film. Of the thirty-six microphones placed along the motorcade route, the one that recorded the sequence of impulses that matched the third impulse on the 1963 dispatch tape was farther along the route than the one that recorded the impulses that matched the second impulse on the dispatch tape. The locations of the microphones were consistent with the distance a motorcycle traveling at about 11 mph would cover in the elapsed time between impulses on the dispatch tape. This relationship between the location of the microphones and the time between impulses was consistent for the four impulses on the dispatch tape, a strong sign that the impulses on the 1963 dispatch tape were recorded by a transmitter on a motorcycle or other vehicle proceeding along the motorcade route. Applying a statistical formula, Barger estimated that since the microphones clustered around a line representing the speed of the motorcade, there was a 99 percent probability that the Dallas police dispatch tape did, in fact, contain impulses transmitted by a microphone in the motorcade in Dealey Plaza during the assassination.[11]

The most interesting match occurred near the last sequence in order. This match was most strongly correlated to a shot originating from the grassy knoll. However, because the microphones recording this signal were almost certainly not positioned at the exact spot of the motorcycle that made the original Dictabelt recording, Barger could only assign a confidence level of 50/50 to this result.[12]

Clearly, to establish the validity of this match would require further analysis.

The committee then turned to Weiss and Aschkenasy to extend the analysis of the BBN team. Weiss and Aschkenasy developed an analytical extension of Barger's analysis to help them improve the probability estimate using the technique of sonar modeling. A computer simulation would not be constrained by the locations of the microphones, but, in principle, could interpolate, or calculate the best positional and acoustical match, between them. Today computer simulations are used in many modern fields of research and are vitally important in fields as diverse as shock physics, solid state science, radiation damage of materials, astrophysics, and molecular biology.

HOUSE SELECT COMMITTEE ON ASSASSINATIONS

Weiss and Aschkenasy, specialists in sonar applications with the Computer Sciences Department of the Queens College of the City University of New York, examined Dealey Plaza carefully to determine which structures were most likely to have caused the echoes recorded by the microphone in the acoustical reconstruction that had exhibited a match to the shot from the grassy knoll. They verified and refined their identifications of echo-producing structures by examining the results of the 1978 reconstruction. This approach allowed them to look for matches in the data within a 1 ms correlation, as opposed to the 6 ms correlation used in the BBN work. Matches at this level of temporal precision substantially reduced the possibility that a match could occur as a result of random noise.

In Dealey Plaza, echoes from gunshot test patterns arrive in two discrete clusters, differing in time by about 190 ms. Echoes originating from structures along Elm Street arrive within 85 ms, while echoes from structures farther back on Houston Street arrive in the last 95 ms of a typical 370 ms duration test pattern. In addition, a "muzzle blast" is usually prominent at the beginning of a gunshot acoustical pattern, while an N-wave (a shock wave traveling faster than the speed of sound due to the rifle bullet exceeding the sound barrier) arrives prior to the

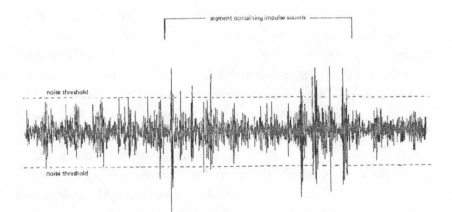

Figure 13: Oscilloscope waveform identified as the grassy knoll shot. Note the appearance of two distinct pulses of data in the segment containing the impulse sounds. The dotted lines indicate the threshold that impulses must exceed to be counted as significant data above the background noise.

muzzle blast. The waveform identified as the grassy knoll shot is shown in figure 13.[13] The presence of an N-wave in this waveform was consistent with the acoustical signature of a supersonic rifle bullet.

Weiss and Aschkenasy were also able to incorporate the movement of the motorcycle in their analytical model. For instance, over the time period of the 370 ms gunshot acoustical pattern recording, the motorcycle would have moved about five feet. Therefore, corrections had to be made to account for a moving microphone during the actual 1963 events. They also took into account muting and distortion of the acoustical signature due to the windshield of the motorcycle.[14] When they considered this in their analysis, they found that the earlier shots determined by BBN to be from the Texas School Book Depository showed this distortion, while the acoustical sequence that matched the test shot from the grassy knoll did not exhibit this effect.

This is precisely the result expected for a microphone mounted on a motorcycle moving in Kennedy's motorcade along Houston Street facing the depository, where the sound of the shot from this location would pass through the windshield, but moving sideways to the direction of the grassy knoll so that a shot from this location would not need to pass through the windshield to reach the microphone. This lent further credence to the validity of their analysis. Weiss and Aschkenasy concluded that their detailed computer simulation refinements of the original BBN analysis allowed them to assign a confidence level of 95 percent to Dr. Barger's identification of one of the acoustic patterns from the original 1963 Dictabelt recording as a rifle shot originating from the grassy knoll.[15]

Photographs taken shortly before the assassination, however, indicated that Officer D. B. McLain was on Houston Street heading toward Elm as the presidential limousine was turning on to Elm in front of the Texas School Book Depository. At the time of the assassination, therefore, he would have been in the approximate position of the transmitting microphone indicated by the acoustical analysis. Officer McLain testified to the committee that he was the officer in the photographs taken of the motorcade on Main and Houston streets, and that at the time of the assassination he would have been in the approximate position of the open microphone near the corner of Houston and Elm, indicated by the acoustical analysis. He did not recall using his radio or what channel it was tuned to on that day, although he did state that it usually was tuned to channel 1. He stated that the button on his transmitter receiver often got stuck in the "on" position when he was unaware of it, but he did not know if it was stuck open during the time of the motorcade. According to a transcript of channel 2 transmissions, approximately 3.5 minutes after the assassination, Dallas Police Department dispatcher Gerald D. Henslee stated that an unknown motorcycle on Stemmons Freeway appeared to have its microphone switch stuck open on channel 1.[16]

As a result of the acoustical analysis by the two groups, and the evidence that a motorcycle in the motorcade was in the correct position in Dealey Plaza with an open microphone on channel 1 consistent with this analysis, the House Select Committee on Assassinations resolved that Kennedy was likely shot at by a rifle located behind the west side picket fence on the grassy knoll as part of a "probable" conspiracy to assassinate him. The committee believed that at least two gunmen had fired at Kennedy in Dealey Plaza.[17]

In the introduction to the committee's report, Chief Counsel Robert Blakey stated: "Realizing that there would be an opportunity for others to fill in the details—that there might be indictments and trials as a result of future investigations—we decided to present an understated case. We chose a cautious approach."[18] The committee recommended that the Justice Department attempt to determine exactly who was behind the conspiracies to assassinate both President Kennedy and Martin Luther King Jr.

However, the Justice Department chose not to follow up on the recommendations of the HSCA. Instead, it decided to challenge the committee's conclusions. The acoustic results that the committee had relied on were subsequently called into question when the Justice Department commissioned a study by the National Research Council (NRC) in 1980, an organization for whom I worked as a postdoctoral research associate from 1988 to 1990. The NRC group was headed by Professor Norman Ramsey at Harvard University, a distinguished scientist. This team studied the data analysis from the BBN and Weiss and Aschkenasy teams but did not perform additional experiments. The NRC then produced a report of their findings as an independent review panel.[19]

The NRC report cast doubt on the statistics used in the prior acoustic analysis and argued that a comparison of crosstalk exhibited on a second recording taken from a different microphone indicated that the time period in which the shots appeared on the Dictabelt recording was not during the time of the assassination. The basic contention of the NRC analysis was that the statistical calculations performed by the HSCA scientists did not properly take all the analytical factors into account. All models are constructed by fitting raw data to mathematical functions. To fit these functions, adjustable parameters are used. If there are enough parameters, it is possible to fit anything. With seven adjustable parameters, it is possible to "fit a bear"; almost any data set can be modeled by a sixth-order polynomial. However, it wouldn't mean much. It's best to fit data using the least possible adjustable parameters. This means that the right model must first be found. This is similar to Kepler's work on planetary orbits. He fit the data for the orbits to the most economical function possible, the ellipse. He would have liked to have fit the data to a circle, but the circle didn't have the flexibility to accommodate the true paths of the orbits. The ellipse turned out to be the optimal function, with the minimal numbers of free-floating parameters to model the data. Anything more would be overkill; anything less would have been inadequate.

When the BBN group compared their data sets and worked up a statistical level of significance, they had a number of parameters to work with. In particular, they could select the microphone(s) with the best match to the data, as well as the shooter location and target. But the NRC report argued that there were other parameters to consider as well, including additional degrees of freedom for the shooter location on the grassy knoll, as well as the air temperature, which would have affected the sound velocity and hence the waveforms in the Dealey Plaza acoustical reconstruction of the assassination. According to the NRC, because BBN did not take these parameters into account when they calculated their statistical significance of a match to be 95 percent, the statistical significance of their conclusions was erroneous. Correcting for these errors, the NRC calculated the significance to be only 78 percent. This value was too low to conclude statistically that the sounds on the tape were in fact sounds of gunfire in Dealey Plaza during the assassination.[20]

The second salient issue raised by the NRC, a branch of the National Academy of Sciences, was that crosstalk with a second recording synced the time of the channel 1 recording to a different period than the assassination. The phrase "Hold everything secure," first noticed by rock drummer Steve Barber in 1979, appeared on both tapes, but on the channel 1 recording it appeared just after the last sounds of purported gunfire. Since this phrase was known to be uttered by Dallas County Sheriff Bill Decker one minute after the assassination, the time of appearance of this phrase on the channel 1 recording meant that the eleven-second period in which the shots were recorded occurred after the assassination had already taken place. Thus, the NRC report argued, the sound impulses identified as gunshots must have been something else entirely.[21]

The third significant point raised by the NRC report was that the computer simulations didn't look at all the data but instead focused only on the shot believed to have originated from the grassy knoll. If all the supposed matches to the test data were subjected to the same computer analysis, then it would facilitate evaluation of the legitimacy of the analytical technique and possibly add support to the conclusions of the original study. However, in their conclusion, they argued that while analysis of all the matches was desirable, it probably wasn't worth the effort or expense because the matching sounds weren't coincident with the time of the assassination.

There is an old joke in the academic community. A scientist, a mathematician, and a statistician are asked to solve a problem. They are asked, "What does one plus one equal?" The scientist steps to the blackboard. He begins to write. "Let's see now . . . one electron plus one electron yields two electrons. So, the answer is two." The mathematician steps up. He, too, writes on the blackboard. "OK, I've got one plus epsilon, added to one plus epsilon. That gives me the value two plus two epsilon, but as epsilon goes asymptotically to zero, the answer converges to two identically." Then the statistician gets up. He consults his tables, then his graphs, then his charts. He pulls out his calculator. He looks up and says, "What do you want it to be?"

The ability to shoot down other research groups is a valuable commodity in science, and often crucial when confronted with the reality of funding competition. Casting doubt isn't hard. But what convinces scientists most of all is lab-to-lab reproducibility. Consistency between independent experimental results and consistency with theory, over time, usually carries the day in science.

For instance, over the years many tests of general relativity have been performed. As technologies improve, ever-more-accurate tests of the theory are made by ever-more-complex types of experimental apparatus. So far, the predictions of general relativity have always been found to agree with the experimental results. Many competing theories have been formulated, adding additional "critical" terms to the theory, but these are never needed because the original theory always falls within the error bars of each and every experiment. It is this consistency between experiments and theory, and most important, between widely different experimental designs and types of apparatus, that led the scientific community to accept general relativity as truth. By the same token, it is the extraordinary consistency between the timing of events in the acoustic analysis and the Zapruder film analysis that makes a convincing case for the validity of the acoustical conclusions.

While it is always desirable to complete an analysis to the fullest possible extent that the data allow, sometimes the realities of time and funding prohibit it. I once attended a program review for academicians who were presenting their work to a funding review board. One professor, after relentless criticism from his colleagues in attendance, replied, "OK, give me infinite time and infinite funding, and I will do the perfect experiment." In the real world, the perfect experiment doesn't exist. Funding limitations are often the constraint in science that impedes progress. Just ask the US high-energy physics community how progress in their field was set back when the Super Conducting Supercollider, then under construction in Texas, was canceled in the early 1990s. It would have been terrific if the computer simulations could have been extended to all the data, but this doesn't negate the significance of the work that the acoustic experts did do.

The National Research Council admitted that its calculations of the relevant probabilities may have been "unduly conservative."[22] In fact, the report uses the word "conservative" with regard to the statistical analysis a number of times. It then goes on to state: "Except for the rather conservative analysis above, the data do tend to cast doubt on the hypothesis of random impulse locations."[23] In other words, the sounds on the tape do not appear to be caused by standard random noises, like cars backfiring in Dealey Plaza. This seems to me a lot like an admission that the Bolt, Beranek & Newman, Inc., impulse sequences were indeed nonrandom data, and therefore attributable to something significant. As gunshots were the only reasonable source that matched the acoustic signatures, it would appear that the NRC doubted its own conclusions.

With regard to the crosstalk synchronization, the NRC report states, in an appendix discussing this analysis, that:

> The analyses in Appendixes B-1 and B-2 [of the crosstalk] may be subject to some criticism. A certain amount of subjectivity derives from the fact that the first observer was looking at the sound spectrograms from both channels while he marked points on Channel I.... However, this experiment was supplemented by several variations that derived similar results. Some of these were more careful to avoid the subjectivity and to reduce considerably the dependence aspects of the experiment presented here. These are not reported in detail, because they were carried out using xerographic copies of photographs using several scales, and relatively crude measuring instruments (graph paper in place of rulers).[24]

This is an admission that the first technique the NRC used to confirm the crosstalk between the channels and to determine that the impulses of interest were not concurrent with the assassination was highly subjective and therefore subject to doubt. Moreover, the method they used to confirm the first technique was so low-tech that they didn't even want to describe it. I could speculate that they employed the zoom function on the photocopier; however, precisely how they did this analysis was deliberately left vague and so there is no hope of reproducing their work. In science, if other scientists are to take conclusions seriously, it is essential to describe the methodology of *what was done to reach them* in detail. Otherwise, the work doesn't have validity, because no one will be able to reproduce the results.

Dr. Barger himself responded to the issue of the crosstalk synchronization in a letter to G. Robert Blakey in 1983, after release of the NRC report, concluding that the NRC synchronization was doubtful and that the original BBN conclusions remained valid:

> [Steve] Barber discovered a very weak spoken phrase on the DPD [Dallas Police Department] Dictabelt recording that is heard at about the time of the sound impulses we concluded were probably caused by the fourth shot. The NAS Committee has shown to our satisfaction that this phrase has the same origin as the same phrase heard also on the Audiograph recording. The Audiograph recording was originally made from the channel 2 radio. The common phrase is heard on channel 2 about a minute after the assassination would appear, from the context, to have taken place. Therefore, it would seem . . . that the sounds that we connected with gunfire were made about a minute after the assassination shots were fired. Upon reading the NAS report, we did a brief analysis of the Audiograph dub that was made by the NAS Committee and loaned to us by them. We found some enigmatic features of this recording that occur at about the time that individuals react to the assassination. Therefore, we have doubt about the time synchronization of events on that recording, and so we doubt that the Barber hypothesis is proven. The NAS Committee did not examine the several items of evidence that corroborated our original findings, so that we still agree with the House Select Committee on Assassinations conclusion that our findings were corroborated.[25]

In a peer-reviewed publication appearing in the journal *Science & Justice* in 2001 authored by D. B. Thomas of the US Department of Agriculture's Subtropical Agricultural Research Laboratory, the original HSCA findings were rehabilitated.[26] Dr. Thomas re-performed the statistical analysis and found an even higher probability of a real event on the original Dictabelt recording. After a thorough analysis considering all factors, he determined that it was at least 96 percent probable that a second shooter fired at least one shot from the grassy knoll from behind the west side wooden picket fence.

Thomas, who is an expert in statistical analysis, carefully considered all the factors that had gone into the determination of the statistical significance of the matches from the original Dictabelt recordings and the live-fire 1978 reconstruction. He found errors in both the original HSCA analysis, though minor, and major errors in the NRC report. Moreover, the errors in the HSCA analysis were on the conservative side, so that when he recalculated the significance, he came up with a slightly higher level of confidence than the original report, 96 percent, a level of significance sufficient to dismiss random noise as the source of the acoustical matches to gunshots in Dealey Plaza.[27]

The major problem with the NRC analysis, Thomas argued, was that they were considering factors to be free-floating adjustable parameters that were not adjustable in the original analysis, but that were instead tightly constrained. In particular, the air temperature in Dealey Plaza at the time of the assassination was known and used in the computer simulations; it wasn't a parameter that was allowed to vary to match up the data. Further, the rifle location on the picket fence was not variable in two dimensions, but only along the fence, so that its degree of freedom was in one dimension only. In other words, the NRC argued that the BBN group just varied many adjustable parameters until they got a match to the data, a process that would reduce the statistical significance of the BBN results. Therefore, because they mistakenly assumed too many degrees of freedom in the analysis, Thomas argued that the lower estimate of the significance of the acoustical matching calculated by the NRC analysis was therefore erroneous.

TABLE 1: TIME AND LOCATION OF WAVEFORMS MATCHING GUNFIRE IN DEALEY PLAZA. DATA FROM BBN REPORT.

Time of First Impulse (sec)	Corresponding Zapruder Frames	Microphone Number	Rifle Location	Correlation Coefficient
136.2	151–155			all < 0.5
137.7	178–182	2(5)	depository	0.8
		2(5)	depository	0.7
		2(6)	depository	0.8
		2(6)	knoll	0.7
139.2	206–210	2(6)	depository	0.8
		2(6)	depository	0.6
		2(10)	depository	0.6
		3(5)	knoll	0.6
140.3	226–230	2(11)	depository	0.6
144.9	312	3(4)	knoll	0.8
		3(7)	depository	0.7
		3(8)	depository	0.7
145.6	323–327	3(5)	depository	0.8
		3(6)	depository	0.8
		3(8)	depository	0.7

Thomas points out that a study performed in 1992 by Jeffrey Lotz of Failure Analysis Inc. showed Connally's lapel flop out at Zapruder film frame Z224. The sudden movement of the lapel was possibly an indication of a tumbling rifle round exiting his body, or possibly bone fragments ejecting at high velocity. If this analysis and its interpretation are correct, it suggests that Connally was struck at about frame Z224. However, Connally's lapel may have flopped out as a result of a gust of wind. Moreover, the bullet did not exit through the lapel but several inches away from it.[28] Connally himself believed that he was hit between frames Z231 and Z234.[29] Mrs. Connally believed that her husband was struck somewhere between frames Z229 and Z233.[30]

When the sound impulse sequences from the 1963 recording are plotted as a function of time and matched to the locations determined from the 1978 live-fire tests, as shown in table 1, a pattern emerges. If the shot from the grassy knoll is synced to frame Z312 of the Zapruder film, then the correspondence between the timing of the shots exhibited in the film and the timing from the Dictabelt recording is remarkable. The shot timing matches well with a shot from the Texas School Book Depository striking Connally from behind between frames Z229 and Z234, which is consistent with the time the Connallys believed that the governor was struck, given that from two to three frames (2.4 frames) would have been required for the Mannlicher-Carcano bullet to travel the 265 feet from the sixth floor of the book depository to Kennedy's limousine. This means that, according to the testimony of the Connallys, who had both examined the Zapruder film, the shot that struck the governor was fired between frames Z226 and Z231, which perfectly agrees with the time of the shots from the acoustic analysis, as can be seen from the table. A shot taken at Z208 is consistent with a shot fired as Kennedy emerged from under the large oak tree that obscured a shooter's view in the sixth-floor window of the depository. Moreover, the CIA concluded, through analysis of the Zapruder film, that the first shot must have been fired at Kennedy just prior to frame Z190,[31] which is consistent with a shot at frame Z182. This level of consistency with the Zapruder film is impressive. It is also consistent with reports of eyewitnesses and Governor Connally himself that he and Kennedy were struck by two separate bullets.

The HSCA further examined the motion of a young girl, dressed in a yellow shirt and red shorts, who can clearly be seen running in the grass along Elm Street with the motorcade beginning at frame Z133. However, by frame Z182, the girl begins to stop. By frame Z189, the girl has completely stopped and can be seen looking back toward the book depository. When this girl was interviewed by the committee, she told them that she stopped because she heard the sound of a gunshot coming from behind her.[32] This would be consistent with a shot fired between frames Z178 and Z182. This matches up well with the timing of the first shot given in table 1 and further buttresses the validity of this synchronization of the Zapruder film with the acoustic record.

Table 1 shows that some shots were spaced very closely together in time. The pairs of shots at 139.2 and 140.3 seconds and at 144.9 and

145.6 seconds occurred within a second of one another. This is too fast for a single shooter of the Mannlicher-Carcano rifle to have fired, even if the HSCA value of 1.66 seconds between shots for the open-iron gun sight is taken as the correct value for the firing time between shots for the shooter in the depository.

The HSCA did not consider this specific synchronization because they believed that the shot from the grassy knoll missed and therefore did not cause the head wound apparent in frame Z313. The committee believed that the medical evidence indicated a shot entering from the rear of the head, leaving a "wake" pattern of fragments in the brain and exiting through the side. However, from a medical standpoint, a shot originating from the grassy knoll and striking Kennedy's head from the right front side was possible, even probable. This was pointed out to the committee in a memo by Dr. Cyril Wecht: "A soft nose bullet, of some other type of frangible ammunition, that would have disintegrated upon impact could have struck the right side of JFK's head in the parietal region. Since this kind of ammunition would not have penetrated deeply into the brain, there would be no evidence of damage to the left cerebral hemisphere, nor would the fragments of such a missile be deposited in the left side of the brain. Also, there would not be a separate exit wound if this kind of ammunition had been used."[33]

In his paper, D. B. Thomas further addressed the NRC report's argument that a syncing of the phrase "Hold everything secure," seemingly appearing on the channel 2 tape, rules out the possibility that the eleven-second period with the impulses of interest could have been contemporaneous with the assassination. Chief Decker's words occur one half second after the sound of the last putative rifle shot on the channel 1 tape. The NRC presumed that the presence of a garbled fragment of Chief Decker's channel 2 broadcast on channel 1 meant that acoustic patterns on channel 1 were not recorded at the time of the assassination, since the Decker phrase occurred one minute after Chief Jesse Curry broadcasts, "Go to the hospital" right after the assassination on channel 2. However, a second instance of crosstalk on channel 2 provides an alternative synchronization of the tapes. A patrolman named

Bellah broadcast the following on channel 2, 180 seconds after the Curry transmission: "You want me to hold this traffic on Stemmons until we find out something, or let it go?" This identical broadcast occurs clearly on channel 1, 171 seconds after the purported gunshots. Adjusting for tracking time on the original recording, the elapsed real time would have been 179 seconds. Using the Bellah crosstalk to synchronize the transmissions of the two channels indicates that the putative gunshots were contemporaneous with the instant Kennedy was shot.

One of the crosstalk episodes between police radio bands must have been out of order since the elapsed time between the Curry and Bellah crosstalk broadcasts is 171 seconds apart on channel 1 but only 120 seconds apart on channel 2. The NRC assumed that the channel 2 recorder, which was sound activated, may have stopped for 60 seconds during this interval. However, an alternative explanation is possible. Dr. Barger suggested that the barely audible fragment of Decker's broadcast on channel 1 could be an overdub resulting from the recording needle on the Dictabelt jumping backward in its track. Because regression analysis showed that no time was missing from the relevant section of the channel 2 tape as was erroneously assumed in the NRC work, the fragment of the Decker broadcast appearing on channel 1 is only explainable by the overdub hypothesis. In any case, the eleven-second sequence of impulses on channel 1 resembling gunshots falls neatly into an eighteen-second time interval between two successive Chief Curry broadcasts that bracket the time of the shooting on the channel 2 recording.

Lastly, Thomas argues that even though the NRC panel concluded that the impulse sounds on the original tape were not taken in Dealey Plaza during the time of the assassination, it offered no plausible explanation for the presence of loud impulses on the tape.[34] A vehicle backfire or even a firecracker would not have been loud enough to exhibit the waveforms that were observed. For these reasons, Thomas concluded that the NRC report was wrong. The NRC's contention that the putative impulses on the channel 1 tape were not synchronous with the assassination did not stand up to scrutiny.

FIGURE 14: DICTABELT AND AUDIOGRAPH SEQUENCE

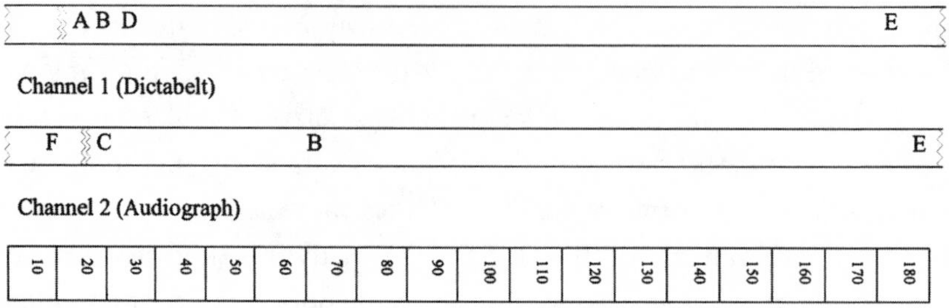

| A B D | | | | | | | | | | | | | | | | E |

Channel 1 (Dictabelt)

| F | C | | | B | | | | | | | | | | | | E |

Channel 2 (Audiograph)

| 10 | 20 | 30 | 40 | 50 | 60 | 70 | 80 | 90 | 100 | 110 | 120 | 130 | 140 | 150 | 160 | 170 | 180 |

A. LAST IMPULSE PATTERN.
B. DECKER: "Hold everything secure." (Garbled on Channel 1.)
C. CURRY: "Go to the hospital."
D. Carillon bell.
E. BELLAH: "You want me to hold this traffic on Stemmons until we find out something, or let it go?"
F. CURRY announces position of motorcade.

Data from NRC report Table C-1

The publication of D. B. Thomas's paper was not the end of the story, however. In 2005, a group led by scientists at the IBM Thomas J. Watson Research Laboratories in New York published a paper in the same journal that Thomas had published in, *Science & Justice*, the journal of the British Forensic Society.[35] The Thomas J. Watson research laboratory is a famous research institution with a world-renowned reputation and staff located in a beautiful area of New York State called Yorktown Heights. The building itself features enormous open curved-glass windows that form the outer wall, so that a gorgeous view of a scenic wooded area is afforded all who walk its halls. This laboratory was one of the first facilities to have luxuries like e-mail and instant messaging back in the 1980s. When I have published papers in the *Physical Review*[36] and in *Physical Review Letters*,[37] those papers were reviewed by staff members at T. J. Watson. Because their work is heavily weighted to semiconductor research in support of IBM's computer technology, the T. J. Watson staff's capabilities and expertise in areas like solid-state physics and materials science is respected worldwide.

However, note that Norman Ramsey of Harvard University, the key member of the original National Academy of Sciences (NAS) committee, was a coauthor on this paper. The new paper essentially rehabilitated the earlier report by the NAS, and concluded that D. B. Thomas was in error in his calculation of the probability that the shot reputedly fired from the grassy knoll was due to random noise and that the original National Academy of Sciences report was accurate in its conclusions that the acoustical signatures in the data were misinterpreted as gunshots.

In science, even the best ideas are not always accepted at first. The theory of continental drift, for instance, first proposed by Alfred Wegener in 1912, was not accepted until the 1950s. Wegener labored for years to demonstrate his theory of continental drift, which asserts that the continents essentially float on large plates held up by the earth's mantle.[38] These plates could therefore move around the earth over long periods of geological time. At one time, all the continents were together in a single giant landmass that he called Pangaea. Although Wegener had assembled a large amount of circumstantial evidence supporting his controversial theory, his views were met mostly by skepticism from the scientific community because he could not provide a mechanism for the phenomena. Scientists could simply not believe that the continents were somehow able to "plow through" the solid oceanic crust.

In 1943 the famous George Gaylord Simpson wrote a fervent rebuttal to Wegener's theory. Simpson's reputation was so great that he influenced most geologists to dismiss Wegener's theory out of hand. However, by the 1950s newly discovered evidence from the nascent field of paleomagnetism supported Wegener's controversial hypothesis. In 1953, rock samples taken from India demonstrated that the country was once in the Southern Hemisphere, as predicted by Wegener. In the 1960s, the discovery of sea floor spreading led to the revival of the theory of continental drift. Today Wegener is considered one of the key pioneers of geology and his theory a major contribution to science. Wegener himself died on an expedition to Greenland in 1930, never having lived to see the acceptance of his theory, which today is considered an established fact taught to grade school students.[39]

Einstein is known for his theories of relativity. But it is a little-known fact that Einstein did not win the Nobel Prize for his work on relativity. He won it in 1921 for the photoelectric effect, a paper that had heralded the beginnings of quantum theory, perhaps the foremost scientific achievement of the twentieth century.[40] This paper had been published in 1905, Einstein's *Anna mirabilis*, the same year in which he had published special relativity. However, in 1921, relativity was still extremely controversial in the scientific community. As a result, the Nobel Prize committee cautioned Einstein not to discuss relativity in his acceptance speech. Instead, Einstein took the podium, thanked the committee, and promptly proceeded to explain his theory of relativity for the audience. He spent almost his entire Nobel acceptance speech talking about the theory of relativity. History, of course, would vindicate him. Even the best theories in science are often controversial at the outset, and in science the outset can last decades.

Another famous example of this delayed acceptance of a theory in science was provided by Samuel Goudsmit and George Uhlenbeck, two young scientists who, in 1925, made an extraordinary discovery.[41] They found that moving electrons would be deflected in a magnetic field in different directions. This implied that electrons had a magnetic moment, which the two young scientists attributed to "spin." If a charged object is spinning or rotating like a top, akin to the rotary motion of Earth's metallic core, it produces a magnetic field. The rotation of Earth's iron-nickel core, for instance, is responsible for Earth's magnetic field, the same field that protects us from charged-particle radiation from the sun, preserves our atmosphere and oceans, and keeps us in good health. Therefore, the pair believed that they had discovered evidence for a key property of electrons, "spin," and wrote up a paper on it for submission to a journal. However, when they discussed their theory with a famous physicist of the day, Hendrik Lorentz, a Nobel laureate whose work had helped lead Einstein to the theory of relativity, Lorentz was skeptical of their ideas. As Uhlenbeck later recalled, Lorentz had told them, "A charge that rotates? Yes, that is very difficult because it causes the self energy of the electron to be wrong."[42]

Lorentz didn't believe that something so small could rotate at the speed required to produce the "spin" that they were postulating.

But Lorentz was thinking "classically." Electron "spin" turns out to be a quantum mechanical entity. At that time, physics was in the midst of a "quantum revolution," a totally new and different way of viewing nature. The discussion with Lorentz scared the two young scientists, so they went to their academic adviser, the famous Paul Ehrenfest, hoping to head off the submission of their paper and so save their careers. They told him, "Don't send it off, because it probably is wrong; it is impossible, one cannot have an electron that rotates at such high speed and has the right moment." And Ehrenfest replied: "It is too late, I have sent it off already."[43] Ehrenfest is reputed to have told them that they were young and so could afford to take some risks with their careers.

But it would turn out the Goudsmit and Uhlenbeck were right. Electrons do have a property called "spin," it does produce a magnetic moment for electrons, and it is essential to the understanding of their behavior. The two young scientists had made a major scientific break-through of extraordinary impact, and fortunately they had done so in spite of the disagreement and objections of more established and "famous" scientists with estimable reputations. We would not understand physics, chemistry, or biology today without their ground-breaking work.

It turns out that "spin" makes the world go round. It is a crucial property of fundamental particles, of not only electrons but also protons, neutrons, pions, partons, and just about any other particle you can think of. Spin is the property that puts the *super* in super string theory. Spin is what makes the electronic structure of atoms possible. If it wasn't for "spin," we wouldn't be here.

The moral of the story is that the established Nobel-winning scientists aren't always right. A great reputation is no proof against being wrong. In general, criticizing a successful experimental scientist, like Dr. Barger, in his area of expertise is a dicey proposition. Someone who does acoustical analysis for a living is not likely to make major mistakes in his field of investigation. Such an individual is familiar with the pitfalls and knows the false alarms and the "fool's gold" in his specific discipline. He is not likely to misidentify a person's voice on a recording as a gunshot. Nor is he likely to mistake one sound for something else. When his results are supported by other expert practitioners in acoustics, then his work becomes doubly difficult to successfully criticize.

Leaving reputations aside and focusing only on the data, who is more likely to be right? The correlations between the acoustic analyses conducted by BBN and Weis and Aschkenasy with the Zapruder film are compelling. The correlation between the matching of the microphone wave patterns and the speed of a motorcycle moving along the JFK motorcade route is compelling. Taken together, these correspondences are overwhelmingly convincing. For six microphones to randomly match up in the correct sequence for a procession along Elm Street would be about 1 chance in 720, because there are $6 \cdot 5 \cdot 4 \cdot 3 \cdot 2$ different possible sequences. This doesn't even take into consideration that the timing of this sequence matched the known velocity of the motorcade at about 11.2 miles per hour. What are the odds of that happening randomly? In principle, the timing sequence of the shot wave forms from the live-fire tests could range from near zero miles per hour to a thousand miles per hour and anything in between, if they occurred as random noise. One could certainly insert a big number here for the total number of possibilities, leaving a very small probability that this would happen randomly. But it isn't necessary.

Matching the time of the appearance of shots on the Zapruder film gives a second, corroborating high level of confidence that the 1963 acoustic records are the records of gunshots in Dealey Plaza at the time of the assassination. Syncing the final head shot from the grassy knoll to frame 312 (we can see on the film that it occurred between 312 and 313), the probability of finding the shot that hit Connally to within five frames (given the inaccuracy and the variation of the speed of the original recording) is about one chance in a hundred, given that there are nearly five-hundred Zapruder film frames. Matching up the first shot to the frames before Kennedy reaches the Stemmons Freeway sign and the second shot to a strike of Kennedy behind the sign is another one chance in a hundred times one chance in a hundred for a one-in-ten-thousand chance for an accidental match. Multiplying this by the probability of getting all four shot origins correct in order is another one chance in sixteen, yielding a one-in-sixteen-million chance that the acoustic analysis could match up the timing and shot sequence in the Zapruder film by chance. These statistics alone, yielding an over one-in-a-million chance of happening randomly, exceed the accepted statistics for identifying a new phenomenon in science. These statistics are more than good enough to announce a major scientific discovery and be accepted by the scientific community.

But when the match of the shots fired to the Zapruder film is coupled with the match-up of the motorcycle moving along the motorcade route, an enormous probability is determined against both sequences being random. Through multiplying the two probabilities together, it is readily established that there is only one chance in eleven billion that both correlations could occur as the result of random noise. This is about the chance of winning the lottery two nights in a row, and by buying a single ticket each time. (If you do win back-to-back lotteries, you still can't dismiss the acoustics data, although in that case you probably won't care then one way or the other.)

So, given the straightforward analysis just performed, when the statistical calculations for the significance of the acoustic analysis are all added up, who is more likely to be right? Is it the acoustic experts who calculate a 96 percent confidence in their results, or the scientists from other fields who argue that something like 70 percent is closer to the truth? Who is more likely to be right, the acoustic and sonar specialists who believe that the sounds of gunshots are apparent on the tapes from Dealey Plaza, or the scientists who argue that this section of the tape wasn't even recorded during the time of the assassination? Given that the chance of randomly matching up the data that we *know* was acquired during the time of the assassination—the Zapruder film and other films of the motorcade—exceeds one in a billion, it is very difficult to believe that the acoustic experts are wrong. Since we know from our comparative analysis that the odds of a false match of the acoustic data to the

Zapruder film record is astronomically small, the weight of probability overwhelmingly favors the conclusion that the acoustic data is a sequence of real rifle fire events produced at the time of the assassination.

Numbers don't lie.

CHAPTER 7

RECLAIMING HISTORY?

O most marvelous of men! though you have eyes to see, you do not perceive; though you have ears to hear, you do not recollect.
—Xenophon, *The Anabasis*, book 3, chapter 1

I had avoided taking physics in High School.
—Vincent Bugliosi, *Reclaiming History:*
The Assassination of President John F. Kennedy

Both Vincent Bugliosi's *Reclaiming History*[1] and Gerald Posner's *Case Closed*[2] have prospered in the marketplace, been hailed by the news media, and been embraced by the critics. Both are anticonspiratorial. Both propose the same models of Kennedy's death. Both espouse the single-bullet theory. Both conclude that only three shots were fired. Both pronounce Lee Harvey Oswald as the sole assassin. Both are wrong.

Gerald Posner, a Wall Street attorney, begins his popular book *Case Closed* with a description of Oswald's arrest and a discussion of his childhood and personal history. It is clear from the outset that Oswald will be the central figure of this work. The biography of Oswald goes on for 230 pages, about half the book, before segueing into a discussion of the assassination and of the Dealey Plaza witnesses.

Posner argues that the witnesses who claimed to have seen smoke or heard shots from the grassy knoll did not, in fact, see smoke but instead saw steam from a pipe,[3] and did not actually hear shots from this location. However, in his thorough and comprehensive book *Crossfire:*

The Plot That Killed Kennedy, Jim Marrs writes, "It has been well established that there was no other natural source of smoke in that area [the railroad yard] that day. FBI reports attempted to show that it may have come from police motorcycles, but none were on the knoll at the time."[4]

One by one, Posner examines the testimony of witnesses recounted in Jim Marrs's book *Crossfire* in an attempt to discredit them. For instance, in the case of Jean Hill, who was standing next to Mary Moorman when Moorman took her famous photograph and who claims that she ran up the grassy knoll to seek Kennedy's assassin, Posner rejoins that pictures taken minutes after the assassination refute this since they show that "Jean Hill is still either sitting or standing next to Mary Moorman."[5] If the pictures were taken minutes after the assassination, she would have had time to go up the hill looking and still have time to return to her friend. In his very next paragraph, however, Posner admits that in her original statement, "Hill said when she got to the grassy knoll, the police were turning people back, so she returned to Mary Moorman."[6]

For grassy knoll witness Ed Hoffman, Posner questions the credibility of his story. Hoffman had claimed to have seen a gunman toss a rifle to a compatriot disguised as a railroad worker who disassembled the weapon, placed it into a railway sack, and walked away. Posner writes, "Even those who support Hoffman's story find it difficult to explain how anyone was able to disassemble the rifle in the rail yard when more than a dozen people ran into that exact location less than a minute after the last shot."[7] A minute is plenty of time to break down a gun, put it in a bag, and head out. In Posner's version of events, Oswald moves several large boxes to stash his rifle, runs to the other side of the building, descends five flights of stairs unnoticed, and appears in the lunch room sipping a Coke all within a ninety-second period before being confronted by a policeman and a supervisor in the depository.[8]

But Posner could not address every witness in Dealey Plaza who believed shots were fired from the grassy knoll. Even if he could successfully raise doubt as to the veracity of some of the witnesses's claims, it doesn't mean that all of them were lying or mistaken. Nevertheless, he concludes that "[c]hasing shadows on the grassy knoll will never substitute for real history."[9] But dismissing the eyewitnesses to that history will?

Posner is a vocal advocate of the single-bullet theory. He places a great deal of faith in the work done by Failure Analysis Associates, a firm that used computer simulations to demonstrate that both Kennedy and Connally were struck by the same bullet.[10] These simulations have since appeared on national television and mainly consist of graphic representations of the two men in the limousine, while a yellow line exhibiting the bullet's trajectory appears to indicate that, given their relative seating positions, both men were in alignment to be struck by the same bullet, causing all seven nonfatal wounds. However, if the input data used in the study is in error, or assumptions made for the simulations are not tied to reality, the result of such a simulation can be meaningless. Computers can generate pretty pictures, but unless their model is validated by comparison with experiments, their output can be at best worthless, at worst misleading. Even though the Failure Analysis computer re-creations look impressive, this work is far from being scientifically persuasive. However, Posner does make one statement with which I can agree, to wit, "The single-bullet theory was not the result of positive evidence that clearly established it but an attempt to create a scenario to fit the facts as the commission determined them."[11] This is indeed what happened.

Posner ends his book by stating, "But for those seeking the truth, the facts are incontrovertible. They can be tested against credible testimony, documents, and the latest scientific advances." I couldn't agree more. But Posner continues, "Lee Harvey Oswald, driven by his own twisted and impenetrable furies, was the only assassin at Dealey Plaza on November 22, 1963. To say otherwise, in light of the overwhelming evidence, is to absolve a man with blood on his hands, and to mock the president he killed."[12] This sounds to me more like *Mind Closed* than *Case Closed*.

Vincent Bugliosi, noted author and former California district attorney famous for his successful prosecution of the Charles Manson case, begins his book with a "proof" that people who believe that

Kennedy was killed as the result of a conspiracy are not thinking intelligently about the assassination.[13] Bugliosi was attending a conference of lawyers, the vast majority of whom did not believe that Oswald acted alone. First he asked for a show of hands of how many people had seen the movie *JFK* or had read a book or an article that rejected the findings of the Warren Commission. About 85 to 90 percent of the people in the room raised their hands. Then he said to them, "I'm sure you will all agree that before you form an intelligent opinion on the matter, you should hear both sides of the issue. As the old West Virginia mountaineer said, 'No matter how thin I make my pancakes, they always have two sides.'" Then he asked how many of them had read the Warren Commission report? Very few hands went up. Bugliosi concluded, "In less than a minute, I had proved my point. The overwhelming majority in the audience had formed an opinion rejecting the findings of the Warren Commission without bothering to read the commission's report."[14]

This is akin to asking a room full of scientists how many have read Aristotle's discourses on physics. Very few hands would go up. Few scientists have ever read Aristotle's work on physics. Why? Because they already know his conclusions are wrong. If his conclusions are wrong, his reasoning must be flawed as well. Every class they have ever taken or textbook they have read on introductory physics or the history of science has explained why Aristotle was wrong. Aristotle believed, for instance, that Earth could not rotate because the center was at infinity and everyone would fly off. Wrong. He believed that an arrow stayed in motion only because a vacuum formed ahead of it and pushed it along. In Aristotelian physics, bodies must experience a force to remain in motion. Wrong. If you dropped an object from the mast of a ship, it would fall back into the ocean because the ship would move underneath it. All wrong. In fact, blind faith in the works of Aristotle had really set back the progress of science in the Middle Ages. It wasn't until people began doing experiments that they realized the deficiencies in Aristotle's natural philosophy. It fell to the likes of people like Galileo and Newton to formulate the correct laws of motion, and these are the laws that scientists learn and practice to the present day. Since we know the correct laws, and they always work, as simple experiments demonstrate, we needn't torment ourselves with the errors and lengthy discourses of Aristotle.

This is not to say that scientists don't read Aristotle. They do. Many scientists interested in philosophy, politics, ethics, drama, or poetry have read Aristotle. There is no question that Aristotle made enormous contributions in many diverse fields, to the point that some scholars consider Aristotle the greatest genius to have ever lived. But few scientists waste their time today reading Aristotle's original writings in physics. We know that his conclusions were wrong because simple experiments demonstrate them to be so, that they set back the history of science, and that his arguments aren't worth our time to review unless we are interested in the historical development of natural philosophy since ancient times.

By the same token, it isn't necessary to read the Warren Commission's report, much less the twenty-six accompanying volumes, to know that it is wrong. All one has to do is look at its conclusions. They are self-contradictory. The commission members could not agree on the single-bullet theory, but the report says that even so, they all believed Oswald acted alone. Sorry, but if the single-bullet theory is wrong, Oswald did not act alone. Even worse, of the seven commission members who signed the report, four of them didn't believe its conclusions. At least Aristotle believed in his own work. The majority of the members of the Warren Commission didn't believe theirs. In addition to Richard Russell, Hale Boggs, and John Cooper, Earl Warren also didn't believe the commission's primary conclusion that Oswald killed Kennedy and acted alone. University of Wisconsin professor of history David Wrone tells us, "From the day he assumed chairmanship of the commission until the day of his death, Earl Warren firmly believed that a Soviet conspiracy had assassinated President John F. Kennedy."[15] Another commission member, Gerald Ford, actually altered the medical evidence to make the data agree with their conclusions. Even the president who commissioned the report, Lyndon B. Johnson, didn't believe its conclusions and said so. If the conclusions are misleading, the report is likely to be so as well. A distorted document like the Warren Commission report will only serve to confuse and misdirect. Why then is it necessary to read a report that was corrupted and dismissed by the very men who produced it to form an intelligent opinion when it is more likely to lead us astray instead of to the truth?

After "proving" conspiracy theorists to be hasty judges, Bugliosi goes on to criticize his anticonspiratorial colleagues. Bugliosi takes to task the two prominent Warren Commission defenders who published before him, Jim Moore and Gerald Posner. Moore authored the work *Conspiracy of One* in 1990.[16] Bugliosi highlights how these authors treated the case of Sylvia Odio, a Dallas witness who claimed to have seen Oswald with two anti-Castro Cuban men at her doorstep seeking funds two months prior to the assassination. Oswald in the company of men who might have a reason to kill Kennedy poses a problem for anticonspiracy theorists who contend that Oswald acted alone. However, Moore dismisses Odio as a witness with a single sentence. Bugliosi writes, "Posner is even worse, writing what can only be characterized as a distortion and misrepresentation of Odio's testimony. In his book he dismisses the accuracy of her identification of Oswald as being the man at her door by quoting highly selective testimony of Odio's.... Although Posner expressly tells his readers that 'Odio could not positively identify Oswald when shown photos during her Warren commission testimony,' this is simply and categorically not true."[17] Bugliosi goes on to say that Posner unjustifiably dismisses other key witnesses like Rose Cherami, a prostitute and heroin addict who, on November 20, 1963, told a physician at Louisiana State Hospital in Jackson, Louisiana, that President Kennedy was going to be killed during his trip to Dallas. "If these examples were isolated they perhaps would not be worth mentioning.... But Posner does this type of thing too much to be ignored." Bugliosi goes on to say, "If his book had been more comprehensive, particularly in the vast area of conspiracy, and, more importantly, had more credibility, the enormous conspiracy community would not have had the ammunition that they have used against him."[18]

Having effectively dispatched his competition in the anticonspiracy arena, Bugliosi lays down the gauntlet. "I can assure the conspiracy theorists who have very effectively savaged Posner in their books that they are going to have a much, much more difficult time with me. As a trial lawyer in front of a jury and an author of true-crime books, credibility has always meant everything to me. My only master and my only mistress are facts and objectivity. I have no others."[19] Since truth is Bugliosi's only objective, I'm certain that he will be interested in the analysis that follows.

Bugliosi's first important task, from a scientific point of view, was to rehabilitate the single-bullet theory. As discussed previously, failure of this theory means a second assassin was present in Dealey Plaza and negates the proposition that Oswald acted alone. Failure of the single-bullet theory therefore means a conspiracy of some kind to kill the president. The central thesis of *Reclaiming History*, that Oswald killed Kennedy and acted alone, fails utterly unless the single-bullet theory is viable.

Bugliosi strongly supports the single-bullet theory. In his book, he presents analyses, conducted for the HSCA in 1978, that demonstrate that the single-bullet theory is the only possible model to explain the evidence. Bugliosi concludes that "the 'single-bullet theory' is an obvious misnomer. Though in its incipient stages it was but a theory, the indisputable evidence is that it is now a proven fact, a wholly supported conclusion.... And no sensible mind that is also informed can plausibly make the case that the bullet that struck President Kennedy in the upper right part of his back did not go on to hit Governor Connally."[20]

In this case, an informed sensible analysis of the facts should readily confirm Bugliosi's contentions. If the proposition is correct, then it should hold up under all forms of analysis. All known facts should support it. All relevant tests should demonstrate its correctness. All lines of analysis should prove its accuracy. So, how does this version of the "single-bullet" theory hold up upon careful scientific scrutiny?

Bugliosi contends that, contrary to the claims of conspiracy theorists, Kennedy and Connally were horizontally aligned in the limousine in such a way as to allow a bullet fired from the sixth-floor sniper's nest to pass through Kennedy and strike Connally. He points to the work of National Aeronautics and Space Administration (NASA) engineer Thomas Canning, who was hired by the HSCA to examine the single-bullet hypothesis. Canning plotted out trajectories for bullets fired from the sixth-floor sniper's nest of the depository, one for the "bullet that caused the president's upper back and neck wounds," and another to test the hypothesis that the bullet that exited the president's neck also "went on to cause the wounds of Governor Connally."[21] The problem with the single-bullet conjecture, however, is not the horizontal alignment of Kennedy and Connally; it is the vertical path that the bullet must take to inflict all seven wounds on the two men. Bugliosi agrees with Canning that "the bullet that pierced the president's upper back and neck went on to strike Governor Connally."[22] According to Canning, the bullet then went on to strike Connally, "proceeding on a downward trajectory of 25 degrees below horizontal."[23]

As can be seen from the autopsy photographs, Kennedy's neck wound was just below his Adam's apple. The Adam's apple is above the top of the shoulders. The upper back is below the top of the shoulders. Canning calculated an angle of incidence for the bullet that struck Kennedy's upper back as 21 degrees downward from the horizontal. A bullet entering Kennedy's back at this angle, at a downward trajectory, would exit the front of Kennedy's body at a point about two inches lower on the front of his chest. But Kennedy's neck wound is above this entrance point since the Adam's apple is above the upper back. Kennedy's shirt shows a bullet hole about 5.5 inches below the top of his collar and 1.75 inches to the right of the center back line.[24] For a typical collar 1.5 inches wide, the bullet must rise 6 inches above its initial trajectory to exit Kennedy's throat just below his Adam's apple. This would correspond to a 30-degree angle *above* the horizontal, for a total deflection of 51 degrees from its initial path, given that the bullet traverses about 7 inches forward to exit Kennedy's body and 4 inches upward to exit through his throat. Ultimately, however, this deflected bullet would then exit Kennedy's throat at a high angle and clip his jaw; it could not then hit Connally in the back below his armpit but could

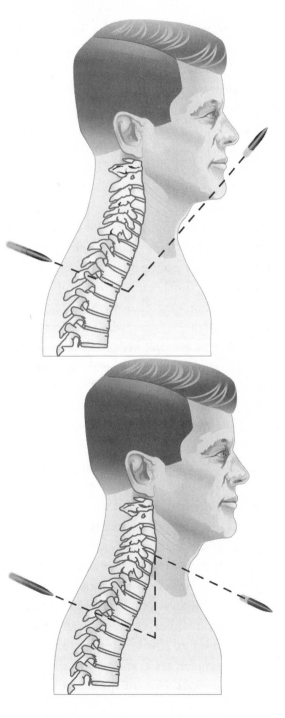

Figure 15: (a) top. Trajectory required by 6.5 mm bullet to enter Kennedy's back and exit his throat.
(b) bottom. Trajectory required by 6.5 mm bullet to enter Kennedy's back, exit his throat, then go on to strike Governor Connally. Courtesy of Linda Alvey-Chambers.

only sail over Connally's head, traveling with such an extreme angle to the vertical. Given that Connally was sitting about three feet in front of Kennedy, the bullet would reach Connally's position twenty-one inches higher than it was when it exited Kennedy's throat. Therefore, for the bullet to go on to hit Connally behind the armpit, regardless of his horizontal positioning, it must then be deflected a second time vertically downward, prior to leaving Kennedy's body, to realign its trajectory with the 25-degree angle of incidence below horizontal required to produce Connally's wounds. See figure 15(a).

For this scenario to hold up, it would oblige the bullet, angling downward as determined at the official autopsy, to reverse direction inside Kennedy's body and reflect backward up from inside his back toward his neck bones, striking a vertebra, reflecting again at a high angle before exiting just below his Adam's apple. This requires a high-angle deflection of greater than 90 degrees with the bullet altering its trajectory from 21 degrees below horizontal, to going backward from the vertical, for a total deflection from its path of something like 120 degrees, backward. Then, to make the overall model hold up, a second extreme-angle deflection would be required by Kennedy's neck bones to redirect the bullet back down again through Kennedy's throat toward Connally's armpit. This would require a second high-angle deflection of 125 degrees back downward to match the direction of Connally's entrance wound. See figure 15 (b).

Deflections at high angles like these for a fast-moving object would necessitate that it encounter a barrier of higher density and thickness. Lead has a density of 11.3 grams per cubic centimeter; human bone has a density of about 2 grams per cubic centimeter. Low-density brittle bones cannot deflect supersonic, high-density hardened metal bullets from 6.5 mm military rifles in the ways required to make this model credible. A military round rebounding backward off Kennedy's bones would be akin to, borrowing a phrase from the famed British physicist Ernest Rutherford, "firing a cannon ball at a piece of tissue paper and having it bounce back at you." If the bullet had enough remaining velocity to punch straight through Connally's back, breaking his ribs then shattering his wrist bone before lodging in his thigh, it had too much velocity to be deflected by bones inside Kennedy's body prior to striking Connally. It would just shatter Kennedy's bones the same way it shattered Connally's and keep right on going along its original path. Unless Kennedy was eating lead for breakfast, or his bones were made of depleted uranium, or he was the bionic man, a full-metal jacket 6.5 mm lead round fired from a military rifle could not bounce around inside his body like a pinball.

In his chapter titled "The Most Famous Home Movie Ever, the 'Magic Bullet,' and the Single Bullet Theory" of *Reclaiming History*, Bugliosi states, "If indeed the [Zapruder] film showed Kennedy and Connally being hit by separate bullets, then the film evidence would be powerful and persuasive. But since we *know* that Kennedy and Connally were not hit by separate bullets, we know, before we even look at the film, that it *cannot* show otherwise."[25]

Let's examine the film carefully, however, even though we "know" that it cannot show that Connally and Kennedy were struck by separate bullets. In science, it's important to keep an open mind and let the data speak for itself. Based on the film, Governor Connally does not appear to exhibit a reaction to being struck until frame Z237, at which point his mouth appears to open. Connally himself testified to the commission that he believed he was hit somewhere between frames Z231 and Z234.[26] Mrs. Connally also examined the film and believed that her husband was struck between Z229 and Z233.[27] In 1992, Jeffrey Lotz of Failure Analysis Associates noticed through computer enhancement of the Zapruder film that Connally's lapel flops out at frame Z224. We pointed this out earlier. He believed that this lapel movement must have been coincident with the exit of a tumbling bullet, or due to bone fragments departing Connally's chest at high velocity. Gerald Posner has argued that this computer analysis establishes Z224 as the frame in which Connally was struck.[28] Under this scenario, Connally exhibits a significantly delayed reaction in frame Z237 to being struck by the bullet thirteen frames earlier (about seven-tenths of a second) at Z224.

It is entirely possible that a gust of wind, for instance, could have caused Governor Connally's lapel to flop out. However, given the Connallys' testimony of when they believed the governor was struck, one can reasonably conclude that frame Z224 was the earliest possible time that the film shows potential evidence of any kind that he was hit by rifle fire. No frames prior to this establish any indication of a reaction from Connally's body, face, or clothing.

By frame Z224, it is apparent that Kennedy has already been hit in the throat. His hands are already raised and moving toward his neck. Human reaction times are typically on the order of a couple hundred milliseconds, or about two-tenths of a second. Nerve conduction travels at speeds much slower than electric currents, for instance, which flow about a foot every nanosecond. Further, nerves must fire in sequence between sensory neurons, control neurons, and muscle-actuating neurons. Neurons fire and recover on millisecond time scales.

A simple test illustrates human reaction times. Have a friend hold a ruler or yardstick in his hand from the top. Place your thumb and forefinger on a fixed known point on the ruler, say, at the ten-inch mark, but don't press them together. Leave enough room for the ruler to slide through. Then ask your friend to drop the ruler without warning, while you attempt to pinch the ruler between your thumb and forefinger to keep it from falling to the floor. If you perform this experiment a number of times, you will find that the ruler falls about eight inches before you can catch it between your thumb and forefinger. This represents your reaction time. Given the rate at which objects fall, 32 feet per second per second, an 8-inch drop corresponds to a reaction time of about a fifth of a second, or about 200 milliseconds.

What is easier to do, pinch your fingers together, or move your hands to your throat? Which of these two movements can be done most quickly? Based on the finger-pinch reaction time test, it is reasonable to argue that 200 milliseconds is about the fastest the human body can react to an outside stimuli. Based on frame Z224, Kennedy must have been struck at least 200 milliseconds prior to this frame, because the movement of his hands to his throat is apparent. This period of time corresponds to about three and a half Zapruder film frames, since his camera ran at 18.3 frames per second. This means that Kennedy must have been struck by frame Z221, or before. In the period of these three frames, a bullet from a Mannlicher-Carcano would travel about 400 feet. The distance between Kennedy and Connally was about only 3 feet. This would mean that for a single bullet to inflict the seven non-

Figure 16: Frame Z224. Connally's lapel can be seen flapping out as Kennedy emerges from behind the sign. Kennedy's hands have clearly moved toward his throat by this time. Zapruder film © 1967 (renewed 1995). The Sixth Floor Museum at Dealey Plaza.

fatal wounds on both men, it would literally, as Kevin Costner's character argues in *JFK*, have to "hang in midair," before striking Connally after exiting from Kennedy's throat. The evidence from examination of the Zapruder film is indeed, therefore, "powerful and persuasive," proof that the single-bullet theory is not viable.

Note that the 200-millisecond reaction time gives a good agreement between the time Governor Connally believes he was struck, between frames Z231 and Z234, and the time at which his face begins to show a reaction, Z237. This reaction time, a fifth of a second, corresponds to about three and a half frames on the Zapruder film. This is a sanity check for our analysis. The reaction time measurement with the ruler is a simple experiment that the reader is encouraged to try at home.

The single-bullet theory ultimately fails *a fortiori* due to another physical law, the conservation of mass. The bullet CE 399 was nearly intact, found on a stretcher in Parkland Hospital and later ballistically matched to Oswald's rifle. This bullet was identified by the Warren Commission as having caused all seven nonfatal wounds to Kennedy and Connally. It weighed 158.6 grains,[29] which is close to the value for a whole bullet prior to firing, 160 to 161 grains. When losses are taken into account due to rifling of the barrel, about 0.5 grains, the weight would be within the average range of standard 6.5 mm rounds, at 159.1 grains.[30] However, substantial bullet fragments were found in Connally's wrist. The Dallas surgeon who worked on Governor Connally, Dr. Robert Shaw, testified to the commission that "more than three grains of metal" were "in the wrist."[31] A metallic fragment was also found on an x-ray of Connally's thigh.[32] A 1968 examination of the x-rays showed the presence of metallic fragments in Kennedy's neck region.[33] Lastly, the FBI weighed the bullet only after first removing two small metallic core samples for testing.[34] It is a basic principle of science that matter can neither be created nor destroyed. The weight of the metallic fragments left in the wounds and removed before sampling easily exceeded the difference between the standard weight of a bullet prior to firing and the bullet recorded as CE 399, which was at most 2 grains. Therefore, the bullet that left the fragments in Connally's wrist could not be CE 399, and therefore an additional bullet must have been fired to inflict the wounds on Kennedy and Connally. This is one more bullet than the "single-bullet" theory allows.

Dr. Shaw made a careful study of Connally's wounds in Dallas. He used a caliper to measure the downward path angle of the bullet through Connally's body, determining it to be 25 degrees downward from horizontal. Dr. Shaw believed that the specific angles associated with Connally's back wound precluded the possibility that the shot that struck him originated from the sixth-floor sniper's nest located in the eastern corner of the depository, but instead must have come from the western end of the building, the side toward the grassy knoll. Because the angle of incidence of Connally's back wound, 25 degrees downward from the horizontal, was higher than the angle of incidence of Kennedy's back wound, 21 degrees, Shaw concluded that it must have come from an elevated location that was closer to the limousine at the time the shots were fired, in order to produce the higher incident trajectory angle from above. The other angles associated with Connally's entry wound supported this contention. Historian Michael Kurtz concludes, "This observation, based on careful medical analysis and precise measurements by a highly qualified thoracic surgeon, destroys the single bullet theory by itself."[35]

The single-bullet theory fails fifteen ways to Tuesday. It fails through analysis of the timing of events on the Zapruder film, it fails when compared to the medical evidence, and it fails under careful scrutiny of the ballistic evidence. Because all available evidence rigorously disproves this scenario, the hypothesis that one bullet inflicted seven wounds on two people and ended up pristine on a stretcher must be rejected. The testimony of the eyewitnesses to the assassination, who all believed that the president and the governor were hit by two separate bullets, is overwhelmingly corroborated by the evidence. Bugliosi's latest attempt to salvage the single-bullet conjecture fails utterly when confronted by the data. It is a classic example of another marvelous theory shot down by a brutal gang of facts.

Once we know that the single-bullet "theory" is wrong, we recognize that there were multiple shooters in Dealey Plaza. That was originally demonstrated in chapter 2 through careful analysis of the Zapruder film. Therefore, we know that the final shot that struck Kennedy, the head shot, need not have come from the rear, but could have potentially originated from another location in Dealey Plaza. Once our minds are allowed to open, it is possible to make headway on the origin of the head shot from the perspective that the preconceived notion that the shot must have come from the rear is simply not true.

With this perspective firmly in mind, consider Bugliosi's argument that the final, fatal shot entered Kennedy's head from the rear, slightly above center, and exited the parietal (side) region at just about his right ear.[36] Copious bullet fragments are clearly depicted in the official x-ray of Kennedy's head in the frontal cortex of his brain. Bugliosi accepts this x-ray as an accurate depiction of the distribution of fragments in Kennedy's brain. In principle, a bullet entering Kennedy's head from the rear could have potentially exited the side of Kennedy's head, depending on its orientation at the moment of impact (in this case, his head would have to have been turned to the left at about a 45-degree angle), or the bullet could have left extensive fragments in the front of his cranium, but it could not have done both. The brain tissue inside Kennedy's skull could not possibly deflect or alter the path of a bullet or cause it to fragment obliquely to its trajectory. Unless Kennedy had a metal plate in his head, the bullet could not have been deflected sideways and would have followed a straight path, one way or the other. Kennedy's brain tissue was simply too soft to redirect it. The scenario Bugliosi advocates breaks the symmetry of the event, placing substantial fragments in one direction (the front of Kennedy's head) and the path of the bullet in another (the right side of his head). Broken symmetries are red flags for flawed analyses.

Bugliosi acknowledges that the Mannlicher-Carcano was a 6.5 mm Italian military rifle, and as such was designed to accord with the requirements of the 1899 Hague Convention, which prohibited the international use of expanding, dum-dum, or frangible bullets that do extraordinary damage to tissue.[37] By international convention, only military full-metal jacketed (FMJ) rounds are permitted to be used by armed forces of the world. These rounds do not deform on impact but instead penetrate the body; they are specifically designed not to produce "superfluous injury." This is what happened to Governor Connally. An FMJ round struck him in the back, penetrated his chest cavity, and exited below his right nipple. FMJ rounds were the only types of rounds ever manufactured for the Mannlicher-Carcano rifle.[38] Oswald, or whoever fired at Kennedy from the Texas School Book Depository, clearly used such bullets as demonstrated by analysis of the bullet fragments found in his limousine, analysis of the intact round CE 399, and the wounds to Governor Connally.

As can be seen from analysis of frame Z312 of the Zapruder film (figure 17), Kennedy's head was turned at about 15 degrees to the axis of his limousine when struck by the fatal round to his head shown in frame Z313, so if a FMJ bullet struck him from directly behind in the back of his head, as would be the case if it had originated from the Texas School Book Depository sixth-floor window, it would have exited through his forehead or eye socket. This was proven by live-fire tests conducted by Dr. Alfred Olivier, an expert on wound ballistics, at Aberdeen Proving Grounds in Maryland in 1964 on skulls filled with gelatin.[39] No doctors reported wounds to Kennedy's face and no such wounds are apparent in the Zapruder film or Kennedy's autopsy photos. Kennedy's head was not turned dramatically to the left, as definitively exhibited in frame Z312, where the side of his face is clearly and fully visible, as would be required if the bullet were to strike from directly behind and exit through the side of his head. Again, the bullet cannot alter its path inside Kennedy's head. This rules out the possibility that Kennedy was struck by a FMJ 6.5 mm round from a Mannlicher-Carcano military rifle, which is similar to the American M-1 rifle used in combat in World War II.

However unlikely, it is not possible to absolutely rule out the possibility that the assassin employed one "custom," frangible round that was essentially handmade, in the volley fired at Kennedy, and that this round happened to be the final one to strike Kennedy at frame Z313.

Figure 17: Z312. Kennedy is seen facing forward, his head only slightly turned to his left just prior to being struck by the fatal head shot. Zapruder film © 1967 (renewed 1995). The Sixth Floor Museum at Dealey Plaza.

However, as we pointed out earlier when discussing the medical evidence, if a 6.5 mm frangible round struck Kennedy in the back of the head, it likely would have blown his head off. This was proven by a live-fire test into the head of an anthropomorphic dummy representing Kennedy conducted by the Discovery Channel in 2008.[40] In this test, a marksman fired at a Kennedy-dummy's head using a .30-caliber frangible round fired from a Winchester rifle. The simulated Kennedy-head literally exploded when struck by this round. The 6.5 mm round is intermediate in size between a small 5.6 mm round and a large 7.62 mm (.30 caliber), but its behavior would be closer to that of the larger bullet. If it didn't actually obliterate Kennedy's head, a frangible 6.5 mm bullet striking from behind would have driven fragments into the forward region of Kennedy's head, producing massive carnage, but in any case could not have exited the side of his head to produce a wound similar to that exhibited on the Zapruder film. It could not make an L-turn inside Kennedy's brain. Therefore, the medical, ballistic, and visual evidence precludes the possibility of a 6.5 mm round striking Kennedy in the back of his head during the assassination. This analysis eliminates the possibility that the Mannlicher-Carcano rifle was the sole murder weapon in Dealey Plaza.

RECLAIMING HISTORY?

In order to account for Kennedy's backward head recoil at the moment of impact, Bugliosi resorts to bizarre and implausible explanations like supersonic jetting and freak instantaneous muscle spasms.[41] Let us examine these hypotheses one by one. The first hypothesis, advanced in the HSCA report, is that Kennedy's head recoiled from frame Z313 to frame Z321 as a result of jetting of tissue from his skull, caused by a 6.5 mm FMJ military rifle round exiting his skull. Bugliosi argues that a high-contrast enhancement of frame Z313 shows graphically that brain matter, bone fragments, and blood are ejecting from the side of Kennedy's head, toward the front.[42] He also explains in a footnote, "The president's head can be seen on the Zapruder film as going not only backward but slightly *leftward.* This movement to the left can be explained by the explosive exiting of the brain matter on the right side of the president's head creating a corresponding propulsive momentum (commonly called a 'jet effect') in the opposite direction, as a rocket recoils in a direction opposite to that in which its jet fuel is ejected."[43]

This explanation of Kennedy's head recoil was originated by physicist Luis Alvarez in 1964. Alvarez became interested in the assassination when he was approached by some of his students at the University of California at Berkeley to offer an opinion on scientific aspects of the assassination.[44] He went on to conduct a series of tests in which he shot melons set on fence posts with ammunition similar to that reputedly used by Oswald. As the rounds exited the melon, a "jetting" effect, produced by internal melon matter exiting with the bullet, caused the melons to recoil backward toward the direction of the shooter.

However, melons are significantly different from human heads. Melons have different densities, different hardness characteristics, and different internal and shell structures. Human heads are attached by bones to a neck and a body, and are not resting unattached on a post or a table. While a relatively light, low-density melon may recoil readily in this fashion, a human head encased in high-density skull bone and attached to vertebrae will not necessarily behave similarly. Historian David Wrone comments in his book that Alvarez was actually paid by the federal government to conduct this study but neglected to mention this in his article reporting the work.[45] In science, funding sources are always acknowledged in unclassified articles and publications produced for public dissemination.

In the case of Kennedy, however, an examination of frame Z313 shows that the matter exiting his head is not all ejecting perpendicularly to the front side of his head. There is an obvious large jet that is moving upward, at a high angle to the vertical. This would contribute little to the left rearward head snap, since only the horizontal component of its momentum in the right forward direction would propel the head backward and to the left. This component would be small because the ejecta is not moving primarily horizontally. Other material appears to be ejecting to the sides of Kennedy's head, and only a relatively small amount appears to be in the direction directly opposite to that of the left rearward head snap along the horizontal, as would be required to propel the head backward. Therefore, this jetting matter would have had to have traveled supersonically to produce the rearward head movement apparent on the Zapruder film. Moreover, Dallas motorcycle patrolman Bobby W. Hargis, who was riding to the left of and behind Kennedy and who can be seen in frame Z313 at the moment of the final shot, stated: "I had got splattered with blood—I was just a little back and left of—just a little back and left of Mrs. Kennedy."[46] Further, a substantial fragment of Kennedy's skull sailed over the back of the limousine, ending up in the street thirty feet behind it. It will be recalled that this was recovered by a medical student and is known as the Harper fragment, which was discussed in chapter 5. Blood spattering a patrolman to the rear of Mrs. Kennedy, who was sitting to the president's left, and skull fragments sailing over the back of Kennedy's limousine are not consistent with the scenario of blood and brain matter jetting to the right front of Kennedy's head as a result of a round exiting in this direction.

However, the major problem with this scenario, the proverbial elephant in the room, is that for the bullet to exit the right side of Kennedy's head and eject blood, tissue, and fragments in such a way so as to produce a jetting effect propelling Kennedy's head back and to the left, that same bullet would have had to enter Kennedy's head from *directly opposite the direction of recoil.* In other words, tracing Kennedy's head recoil backward indicates the path of the bullet that must have blown out the fragments to produce this recoil through jetting. This means that the incident bullet must have struck Kennedy on the other side of his head, coming in from the *left rear* moving toward the right front, *precisely along the path of the induced head recoil.*

This means that the assassin would have had to fire his rifle from a location *on the other side of Dealey Plaza.* He could not have fired this round from the depository because that building was directly behind and to Kennedy's right side at the time of the final shot that struck him. The "jetting" theory places an assassin to Kennedy's left rear on the opposite side of Dealey Plaza, in either the Criminal Courts Building or the County Records Building. This means that whoever fired this shot *could not have been Oswald.* Even if this shot from behind could adequately explain the head recoil through a "jetting effect," it poses an even worse problem for the Oswald-acting-alone theory because it introduces another assassin into the picture. It potentially solves one problem for anticonspiratorialists, only to establish a worse one.

Another concern now arises: a bullet exiting Kennedy's head to produce jetting could not also leave copious fragments in his head. The bullet can't both exit and leave numerous fragments. It can either exit, or leave fragments, not both. Exit means leave. Copious metallic fragments scattered in the head means the bullet fell apart but stayed inside the skull. Since a bullet can't stay and leave, the conjecture that a bullet exited Kennedy's head causing a "jetting effect" suffers from a serious conflict with the medical evidence. The presence of copious fragments in Kennedy's brain, observed and reported by the doctors at the autopsy and confirmed by the x-ray technicians, rules out a bullet exiting Kennedy's head causing a jetting-effect recoil. Note that this conclusion is not based on the official x-rays of Kennedy's skull, which are clearly faked, but on the recollections of doctors and x-ray technicians who examined Kennedy's head wound directly and who saw the real x-rays of his head taken at Bethesda the night of the autopsy. The lack of any wound on the left rear of Kennedy's head, described by neither doctors nor witnesses nor observed on any film records of the assassination, rules out this scenario as well. Case closed.

In November 2008, the Discovery Channel conducted a live-fire test on an anthropomorphic dummy, reconstructing the assassination.[47] This test was well funded, well conducted, well produced, and was aired on national television. The test was conducted under the auspices of assassination researcher Gary Mack, archivist of the Sixth Floor Museum in Dallas. In this test, an expert marksman fired at the anthropomorphic dummy of Kennedy from behind, at the range and angle that would have been used by an assassin shooting from the sixth floor of the depository. The shot was fired at the precise location on the rear of the head that was identified by the House Select Committee on Assassinations in 1978. The result was precisely the same as the HSCA said it was. The bullet blew the top of Kennedy's head off, exactly as depicted in the drawings of the HSCA. But there was *no backward head recoil*. The experts who conducted the test said in advance that they "couldn't necessarily reproduce the backward motion of the head." And they were right. They didn't.

This was a carefully conducted reconstruction of the actual event, as well as it could be reproduced, and thus an experiment that tested the validity of theories advanced regarding the origins of the shot fired that is shown to strike Kennedy at frame Z313. It didn't, however, reproduce any of the features observed on the film. The top of Kennedy's head does not blow off on the film. The top of his head was intact on the Zapruder film. The only visible damage exhibited on the film was to the side of his head. Second, there was no rearward head recoil in the live-fire test. This backward head motion is pronounced on the Zapruder film from frames Z313 to Z321. If the "jetting effect" model had validity, it would have reproduced naturally on the live-fire test on a dummy that simulated Kennedy's internal anatomical structure. Therefore, a 6.5 mm round fired from behind a Kennedy anatomical stand-in did not reproduce the results as depicted on the best-known evidence, the Zapruder film record taken at the time of the assassination. The "jetting" theory doesn't agree with the experiment.

The second theory was that the listless Kennedy experienced a muscle spasm at the precise moment of impact, caused by the bullet passing through his brain. Any theory suggesting muscle spasms or contractions has to deal with the issue of human reaction times. As discussed previously, human reaction times are not infinitely fast. The body does not react instantaneously to insult. When the doctor hits your knee with the rubber hammer, there is a noticeable delay before your leg muscles flex and you lower leg extends upward. This reaction time can be quantified by the simple experiment of dropping a ruler between your thumb and forefinger. As this experiment demonstrates, human reaction times are on the order of hundreds of milliseconds.

Zapruder film frames are snapshots taken at fifty-four-millisecond intervals. On the very next frame after the bullet impact, frame Z314, Kennedy is already moving noticeably backward. On frame Z315, his head has moved even farther to the rear, on the order of several inches. Neck muscles are a very weak muscle group in humans, where the head is designed to rest vertically on the neck, not to be supported against gravity as in the case of nonhuman primates. For this model to be viable, it requires a muscle contraction from the waist. A backward spasm from the waist is a major movement of a large muscle group. It's much harder to do this than it is to pinch your fingers together. Yet, for the theory of the backward muscle spasm to explain the head recoil, Kennedy's large muscle reaction time would have to have been much faster than human reaction times in the finger pinch test, about two hundred milliseconds. Therefore, a simple experiment in human reaction times rules out the possibility that Kennedy's rearward head motion is due to a muscle spasm induced by the incoming bullet striking his brain. Under careful scrutiny, both conjectures to explain the backward head recoil fail systematically.

As Nobel prize–winning physicist Richard Feynman tells us in his extraordinary Messenger Lectures, "If the theory doesn't agree with experiment, it's wrong. It doesn't matter who thought it up, how smart they are, or how many degrees they have, if a theory doesn't agree with experiment, it's wrong."[48] Because it doesn't agree with experiment, the final, fatal head-shot-from-behind theory is *wrong*.

Bugliosi's entire laborious analysis is reminiscent of the ancient practice of adding epicycles, nested circles, to planetary orbits to salvage the Ptolemaic or "earth-centered" model of the solar system. Because the theory of the sun moving around the earth in space just didn't match up the motions of the planets, known since ancient times, the people who held this model had to "tweak" it to get it to agree with astronomical data. Adding "epicycles" to the proposal of circular orbits, making the sun and the planets revolve in small circles superimposed on larger circles allowed them to match up their naked-eye observations, more or less. This meant expanding their nice elegant model of circular orbits into a complicated and ugly monstrosity that was confusing even to diagram, much less think about. The necessity of this remedial fix, however, was obviated by the advance of the Copernican model, which placed the sun at the center of the solar system, effortlessly accounting for the phases of Venus and the retrograde motion of Mars, major unsolved problems for the Ptolemaic "earth-centered" system. After Copernicus, astronomers no longer needed to jump through hoops to explain the motions of the planets. Similarly, the recoil of Kennedy's head is readily explained by placing the direction of the incoming fatal round to his right front side. No hoops are needed with this choice. The impact of the bullet would have immediately caused Kennedy's head to recoil backward. As any good scientist knows, the simplest explanation that fits the facts is almost always the right one.

Placing the fatal head shot to the front and right, however, requires a shooter from the grassy knoll. I find it particularly telling to examine how Bugliosi deals with the fifty witnesses who claimed to have seen or heard shots fired from the grassy knoll.

After the assassination in Dallas, many witnesses, including police officers, ran up the hill of the grassy knoll. Bugliosi begins, "The final point I want to discuss concerns the many Dealey Plaza spectators and members of Dallas law enforcement who ran to the grassy knoll area after the shooting."[49] He goes on to argue that while some may have run there because they thought the fatal shot had come from that location, others, perhaps most, were running there to pursue the assassin. He then suggests that there was only one possible area "where a Dealey Plaza spectator might think an assassin would conceivably fire from...the grassy knoll and pergola area with its walls and heavy foliage...he would know that the parking lot area behind the knoll and pergola would be the only area an escaping assassin could run through."[50]

First, none of the witnesses said that they based their belief that a shot came from the grassy knoll because they deduced that it was the best location for an assassin to be because it provided great cover and a ready-made escape route. They all said they thought that a shot came from that location because they either heard it, saw smoke, or saw an assassin behind the fence (see chapter 3). Second, if the Dealey Plaza witnesses could figure out on the spur of the moment that the grassy knoll was the perfect location for an assassin because of its proximity to Elm Street, its masking cover of fence and foliage, and its unobstructed escape route back through the railroad yard, couldn't the assassin figure that out as well? Couldn't the assassin have reached that same conclusion when he cased Dealey Plaza prior to the arrival of Kennedy's motorcade? In fact, I defy Vincent Bugliosi, or anyone else, to find a better location for an assassin in Dealey Plaza than behind the wooden fence on the grassy knoll. It had cover, it was close to the road, it was in front of Kennedy's position as his motorcade moved toward it, and it had an ideal escape path, the pursuit of which was blocked by the wooden fence and concrete pergola. If the Dealey Plaza witnesses could figure this out on the spur of the moment, certainly an assassin with time to plan could as well.

Bugliosi put five people on death row, Charles Manson, Charles "Tex" Watson, Susan Atkins, Patricia Crenwinkle, and Leslie van Houton, on the testimony of a single witness to whom he had granted immunity, Linda Kasabian. Bugliosi concluded that Oswald fired a shot at General Edwin A. Walker six months before the November 22 assassination based on the testimony of a single witness, Marina Oswald, who changed her testimony many times during the course of the investigation by the Warren Commission. How is it then that Mr. Bugliosi can dismiss out of hand the fifty witnesses who reported seeing smoke, hearing gunshots, or seeing assassins behind the fence on the grassy knoll? Given that one witness is enough to close a capital murder case, how is it then that Mr. Bugliosi believes that the testimony of fifty eye-witnesses isn't sufficient to warrant an investigation? I wish that Mr. Bugliosi would explain this legal calculus to me, because I for one am at a loss to understand it.

Bugliosi acknowledges, correctly, that although Kennedy's head moves backward after frame Z313, it actually goes forward between frames Z312 and Z313. In fact, Bugliosi made good use of this observation in his mock trial in London. While one could argue that this forward movement was due to the impact of a bullet round at frame Z312, the movement could also be accounted for if the driver tapped on his brakes. A deceleration of the limousine could easily cause a listless Kennedy, already wounded at this time, to slump forward. Many witnesses in Dealey Plaza thought that his limousine had actually come to a full stop before the final shot struck him. Some photographs taken at the time show the limousine's brake lights on. However, forward movement from frame Z312 to frame Z313 makes it even harder for the head to reverse direction after Z313, because an incident force would be required to accomplish this, like an incoming supersonic bullet.

Assassination researcher Josiah Thompson has postulated that one bullet struck Kennedy from the rear at frame Z312, and another struck him from the front side in frame Z313.[51] The impact of these bullets, according to Newton's laws of motion, caused the head to surge forward at the first impact from behind then recoil backward when the second bullet struck from the front side. This scenario is possible, however, if the bullet that struck from behind, as argued previously, was not a 6.5 mm round but instead a small 5.6 mm bullet, likely fired from a weapon like the AR-15.[52] Otherwise, the round would have penetrated Kennedy's face, or blown off the top of his skull. Though one may argue it would be improbable for two bullets fired from separate rifles to arrive almost simultaneously, it wasn't at all unlikely if the two assassins were coordinating their shots using a walkie-talkie, or had a pre-arranged spot on Elm Street such that when Kennedy's limousine reached that point, they would both fire. In any case, if this is the correct explanation it requires yet another rifle in Dealey Plaza. Obviously, a plurality of rifles rules out Oswald as the only shooter at the time of the assassination.

Bugliosi is convinced that Oswald, however, was Kennedy's sole assassin and says so repeatedly throughout his book. But it is scientifically very improbable that Oswald fired a rifle that day. After his arrest, Oswald was given a paraffin wax test of his hands and right cheek to determine if they showed the presence of nitrates. This test consisted of pouring hot wax onto his hands and cheek and removing the wax cast after hardening. This process fixes nitrate compounds present on the skin into the wax. A bullet is propelled by a ball or tubular powder that ignites inside the cartridge, forcing the round out of the cartridge and down the barrel of the rifle. Ball powders are primarily made from nitrocellulose, which is a nitrate compound. When this nitrogen-carbon-oxygen compound reacts, it goes from a solid form to a gaseous form, producing compounds like nitrous oxide gas, which is another nitrate. Further, since explosives and propellants never react completely, there is always some residue in the output gases of the original solid nitrate compounds that comprised the ball powder prior to firing the weapon. When a rifle is fired, these hot nitrate gases blow back from the firing chamber onto the shooter's face. Even if the rifle's firing chamber were perfectly sealed, as Bugliosi argues, the hot gases would escape as soon as the firing bolt mechanism was operated to eject the spent cartridge and chamber the next round.

Oswald's paraffin showed that no nitrates were present on his cheek, consistent with someone who had not fired a rifle, but did show that nitrates were present on his hands. Since the paraffin wax seeps down into the pores, it is a very sensitive test. Even washing one's face prior to the test will not remove all presence of nitrates. As someone who has worked extensively with ball powders, I can tell you that reacted powders have a very distinctive odor that is difficult to get out of your skin and clothes. The presence of nitrates on Oswald's hands may indicate that he had fired a revolver, for instance (he was accused of shooting

Officer Tippet on the same day as the assassination), however, nitrates could also have gotten on his hands from other sources, such as paper or ink. The absence of nitrates on his cheek is court-admissible evidence; however, that he had not fired a rifle that day.

Following up on the paraffin results, long-time assassination researcher Harold Weisberg and his attorney, James Lesar, obtained the raw data from Oswald's paraffin wax test through the Freedom of Information Act.[53] After their analysis in Dallas, the paraffin casts had been sent to the FBI Laboratory for a more sensitive spectroscopic analysis. The FBI found the same negative result for the cast of Oswald's cheek as that obtained in Dallas. The cast was then submitted to a third analysis at the Atomic Energy Commission at Oak Ridge National Laboratories in Tennessee, where they performed neutron activation analysis using a nuclear reactor. In this extremely sensitive technique, a sample is irradiated, and the resulting radioisotopes identified through spectroscopy of the radiation, typically gamma rays, that they subsequently emit. The precise energy of this radiation is characteristic of a given isotope and therefore of the element that produced it. These scientists reported that Oswald's cheek casting "could not be specifically associated with rifle, nor could the hand samples be chemically linked to revolver cartridges."[54]

Concomitant with these tests, the FBI had also conducted a control study where seven different men fired the Mannlicher-Carcano rifle and subsequently had paraffin casts taken of their cheeks. Analysis of these casts using neutron activation analysis showed that, unlike Oswald, all of the men who had fired shots in the control study exhibited the presence of nitrates on their cheeks.[55] Therefore, this controlled experimental analysis demonstrated that Oswald had not fired a rifle on November 22, 1963, the day of the assassination and his arrest.

This, of course, explains why Oswald was seen only ninety seconds after the assassination by Officer Marion Baker and building manager Roy Truly, sipping a Coke in the depository lunch room and "didn't seem to be excited or overly afraid."[56] If indeed Oswald could have shot the president, stashed the murder weapon carefully in the back behind heavy boxes, run to the other side of the building, descend five flights of stairs, and appear calm and breathing normally, then he must have been Superman. This explains why Oswald was not seen or heard on the stairs by other employees in the depository who were on the fifth and fourth floors at the time of the assassination and who used the stairs themselves to descend after the events transpired in Dealey Plaza.[57] This also explains why Oswald had no escape plan. It's hard to imagine that someone could plot and carry out the assassination of a president so carefully and methodically, succeed spectacularly, yet plan to escape the scene using only public transportation. When he was arrested, he had $13.80 on him, not enough money to get out of the city.[58] He had to go back to his boarding house to retrieve his revolver. For someone planning for months to assassinate the president, it seems odd indeed that he would not have made the effort to learn to drive a car so as to be able to flee the area. It's especially strange that he would leave the cash he needed for his escape with his wife the morning of the assassination, and even more puzzling that he would bring his rifle yet leave his handgun at home and have to go back for it.

Oswald was not stupid. His taped appearances for his "Fair Play for Cuba" committee and his statements after his arrest indicate he was an intelligent, articulate person. He was an ex-marine. Had he in fact shot the president that day, he would have stashed his rifle quickly, left the building immediately, drove by car to a safe house, and stayed there. He then could have attempted to make his way back into Mexico when the search for Kennedy's killer abated, typically at the point when another candidate suspect was arrested. The Dallas police would have been under intense pressure to make some arrest quickly. Oswald could have rented a car using his alias, Alek James Hidell, a pseudonym unknown to authorities until after they had arrested him.[59]

In his book and at his mock London trial, Bugliosi struggled mightily to find a motive for Oswald even though Bugliosi excelled at finding motives. He had even managed to find the motive for the bizarre Manson murders, unbelievable as it was, and persuade a jury of it. It was indeed no small accomplishment to convince twelve men and women that Manson had ordered some of the most heinous murders in California State history in an attempt to touch off a class-race war he called "Helter Skelter," a delusion inspired in his mind by the songs of the Beatles and the book of Revelation. Yet the best motive Bugliosi could come up with for Oswald was that he was a loner and a loser, trying to become famous, to make a name for himself. Bugliosi writes, "He [Oswald] was a lonely, frustrated, and beaten-down loser who felt alienated from society…one who irrationally viewed himself in a historical light, having visions of grandeur and changing the world; one whose political ideology consumed his daily life."[60]

But if that had been Oswald's motive, why didn't he confess when he had the chance? Once he was arrested, Oswald's chances of escaping justice, if he was guilty, were nil. Oswald, with an IQ of 118, certainly could have realized this. Why then not confess and take his place in history? "Yeah, I did it. I was upset about his Cuban policies, and I decided to take matters into my own hands," or whatever his reason was. There was absolutely no downside to confessing at that point. Why not ensure his own fame and historical significance with a pronouncement of his guilt and his rationale? Oswald had supposedly single-handedly altered the course of world history, why then didn't he make a political statement and accept responsibility for his actions on the world stage given that his goal was to become famous, to be somebody, to make a difference in the world, and to be remembered for idealism and his ideology? Instead, when questioned by reporters after his arrest, he said only that he "didn't shoot anybody," asked for legal representation, and stated, "I'm just a patsy."

The results of the paraffin wax test of his cheek indicate instead that Oswald was exactly what he said he was, "a patsy." Oswald's relationships and level of knowledge of the events in Dealey Plaza may remain a mystery, but he wasn't the shooter. Bugliosi says that he has tried cases with less evidence than that against Oswald. I find this frightening.

But even in the extremely unlikely event that Oswald was one of Kennedy's assassins, the evidence that he did not act alone is overwhelming. As in the Manson case, the case that secured Bugliosi's fame and reputation, could Bugliosi have considered himself successful if he had won convictions for the perpetrators of the Tate–La Bianca murderers but failed to obtain a conviction of the mastermind, Charles Manson? Could Bugliosi have deemed himself an effective public servant if Charles Manson were still running around loose in California? Would it not have been a terrible failure, with likely horrific consequences, had he not been successful in finding and convicting the architect behind some of the most heinous crimes in California history? Even his anticonspiratorial colleague Gerald Posner believes that investigative efforts should be launched to determine if Oswald was acting in concert with associates, and he has stated so publicly. As in the Manson case, wouldn't the greater failure be not to make an attempt to look beyond the crime's direct perpetrator(s) to find those who may have planned it?

A careful examination of Bugliosi's arguments reveals serious flaws in logic. It isn't just a lack of scientific understanding that hinders his analyses; his arguments suffer from a paucity of logical reasoning. As an example, consider the conclusions he draws from the questioning of Dr. Cyril Wecht, a leading forensic authority, conspiracy advocate, and vocal critic of the official autopsy of Kennedy. Bugliosi asks him, "Doctor, you always said you believed there was only one gun firing from the president's right. You never suggested two gunmen there, is that correct?" Wecht answers yes. "But if the shot that you originally thought may have struck the president in the throat is not a frangible bullet, yet the synchronized bullet that you say may have struck the president on the right side of his head may have been, you have to have two gunmen on the president's right side, one firing frangible bullets, the other regular ones. You agree this is pretty far-fetched, don't you?" asks Bugliosi. "I agree that would not be likely at all," Dr. Wecht acknowledged.[61]

Bugliosi goes on to conclude from this exchange that "there is no credible evidence whatsoever that any shots were fired from the president's right side or right front (grassy knoll), and the selection of the general area around the knoll as the site from which to shoot the president makes absolutely no sense at all—the conspiracy theorists' leading medical forensic expert cannot even *hypothesize* a shooting from the right side or right front that is intellectually sustainable."[62]

In the first place, this conclusion flatly contradicts the argument that he invoked to explain why the people ran up the grassy knoll to search for the assassin; it was because they were running toward what they thought would be the logical place from which the assassin could fire and escape. In the second place, it isn't necessary for there to have been two shooters standing behind the fence on the grassy knoll. Bugliosi believes that Oswald fired three bullets at Kennedy; why couldn't a rifleman on the grassy knoll fire two at him? Is there some law of nature that states that an assassin can only use one kind of ammunition? Couldn't he just as easily load a frangible bullet and a nonfrangible into his magazine as two frangibles or two regular, hardened rounds? Couldn't he load and fire a regular round, then load and fire a frangible one? Of course he could. And even if there were two shooters on the grassy knoll, so what? Why couldn't there be two shooters on the grassy knoll conspiring to kill Kennedy? Is there some immutable law of the universe that says only one person can be shooting at a target at a time? There was plenty of room along the fence for more than one assassin. There was even elbow room. Is the concept of two shooters standing next to each other behind the wooden fence in Dealey Plaza really intellectually unsustainable? If it were, then the marksmen who carried out the 1978 live-fire test for the HSCA couldn't have done so. According to Bugliosi, it is intellectually unsustainable that they could have been standing next to each while firing the shots at targets in Dealey Plaza.

Because they lead to absurd logical conundrums, arguments like this carry zero weight, even though Bugliosi considers them definitive. Unfortunately, *Reclaiming History* is replete with these types of flawed analyses. A thousand zeros still add up to zero.

Bugliosi labors under the misguided belief that lawyerly tricks and the ability to steer a witness to a desired statement under cross-examination is a substitute for truth. In this case, the Socratic method fails. As the above example demonstrates, arguments that he considers definitive contain serious and unsustainable flaws yet lead him to make sweeping and erroneous conclusions like negating the possibility of a shot from the grassy knoll. Though a brilliant prosecutor and author of many outstanding nonfiction books, Vincent Bugliosi advances and relies on arguments in *Reclaiming History* that sadly suffer from serious logical inconsistencies. This is the heartbreaking circumstance that has led Mr. Bugliosi to devote twenty years of his life to a flawed and hopeless cause, the rehabilitation of the Warren Commission findings and the establishment of Oswald as the lone assassin of John F. Kennedy.

In the same manner that DNA analysis solves decades-old cases, proper application of scientific laws, principles, and methods reveals Posner's and Bugliosi's books as well as the Warren Commission's report for what they are: obfuscations, mountains of pages that simply don't stand up to scientific scrutiny.

For all *Reclaiming History*'s failings, it does have one redeeming value; it contains an absolutely terrific joke, which I recount here: The conspiracy buffs, who believe that just about everyone was involved in the cover-up of the assassination of JFK, are all lined up in front of God at the end of time asking him, "Tell us, God, who really killed President Kennedy?" When God replies, "Listen, I'm just going to tell you one time and one time only, and then I want you to forget about this matter—Lee Harvey Oswald killed Kennedy and he acted alone." The buffs, in terrible angst, nudge each other nervously and say, "This is a lot bigger than we thought."[63]

Indeed, it's bigger than we all thought.

CHAPTER 8

THE ZAPRUDER FILM

I must have been in the line of fire.
—Abraham Zapruder

Film records are commonly used in science. They are used to observe high-speed events that happen on time scales too fast for the human eye to see, such as explosions, detonations, and high-velocity impacts. They are routinely used in astronomy to investigate distant galaxies and to determine moving objects against a static background of a field of stars. They are used in radioisotometry to determine the presence of radiation-emitting isotopes. Photography is a standard and valuable technique in the toolbox of the modern experimental scientist.

Although digitization has made significant inroads into scientific analyses, there are a number of areas of science today that still utilize film. Certainly, medical x-rays are still taken on film and then transferred to magnetic or optical storage media. Some project that half the world's disk storage space will soon be for medical images.

There are many specialized films today that detect various forms of radiation. Film badges are mandatorily worn by personnel exposed to ionizing radiation; that is, radiation that does significant damage to cell structure or chemistry. Such radiation can, over a lifetime or even in a single episode of high exposure, do considerable damage to one's health. The degree of exposure of a film badge can tell how much radiation a person is exposed to during the course of their duties. When certain exposure levels are reached in the course of a year, or over a lifetime, then the individual receiving the dose must suspend activities in the radiation areas. This monitoring protects workers from overexposure to radiation, thereby preserving their health. Many early researchers, like Madame Curie, suffered severe health problems as a result of their work with radiation sources before the harmful effects of radioisotopes were known. The current use of film emulsions capable of responding to ionizing radiation is not only used in radiation and radioisotope research but is essential for the health and safety of scientists and medical personnel working in the radiation and radiotherapy fields.

Extreme high-speed photography is also still performed on film. Both streak and high-speed framing cameras utilize film. A high-speed framing camera, capable of recording events at microsecond (millionth of a second) resolution, uses a separate mirror for each frame of the film exposed to capture events that occur at ultra-high speeds. A streak camera looks at deformation at a fixed point and can yield precise information on the rapid movement of an edge as a function of time in response to a shock wave. The derivative of this data then yields velocity of a moving edge or wall.[1] High-order detonations and high-velocity impacts are typically imaged with these extraordinarily fast cameras.

Film records don't lie. They contain an enormous amount of information because they show the time-history of an event. If a picture is worth a thousand words, a film record is worth a thousand still photographs. With a record of change, distortion, and acceleration as a function of time, the laws of motion can come into play. The laws of motion are all couched in terminology involving change as a function of time.

Film records last. They are permanent. They can be examined years later and still show important details. They can always be scrutinized for signs of alteration. It isn't hard to detect if a film has been modified. Typical methods for film modification may involve painting or air brushing. These techniques were pioneered in the early days of motion picture film, and even amateurs can detect them if they look carefully.[2] There are many famous "matte" scenes in *Gone with the Wind* (1939), for instance, where curved ceilings or other mansion parts were literally "painted" on to the film. In George Lucas's classic series *Star Wars*, a large number of matte shots were used. When Lucas digitally remastered the sequel to *The Empire Strikes Back*, over a hundred shots were reprocessed to cover the matte lines apparent in the original picture in the scene involving the attack against the imperial walkers. Because changes made directly to film are noticeable, it is always possible to examine the original film record closely for evidence of these types of alterations.

This is not the case with digitized images. Film and photograph processing on a computer is so sophisticated today that almost any change can be made to any digitized film image on a home computer. Such changes are essentially undetectable. However, a true original film record can always be examined for signs of alteration using paints or other applications, which are detectable by experts.

In 1963, several film records were taken in Dealey Plaza by various sources. These include the Nix film, the Muchmore film, and the Mormon photograph. The most famous of these films, perhaps the most famous short film of all time, was taken by Abraham Zapruder, an amateur photographer and dress manufacturer who lived in Dallas.

Mr. Zapruder had planned on taking a film of Kennedy's motorcade; however, on the morning of the Kennedys' arrival in Dallas, Mr. Zapruder had decided not to bring the camera that he had already loaded with film for the president's visit. When he arrived in his office in the Dal-Tex Building, he was beginning to have second thoughts. The weather had not looked good that morning; it had seemed a poor day for filming. But by midmorning, the sun was shining through. Zapruder could not decide what to do. It didn't seem likely that he would get a chance to see the president in any case, given the extent of the crowds likely to be in Dealey Plaza that day.[3]

But his staff persuaded him otherwise. His secretary, Lillian Rogers, remarked that the crowds would be light in the plaza and told him, "The president didn't come through the neighborhood every day." Finally, Zapruder relented, drove home, picked up his camera, and returned to his office.

Abraham Zapruder arrived in Dealey Plaza a little after noon from his office in the Dal-Tex Building at 50 Elm Street, his Bell & Howell

8 mm Director Series movie camera in hand. The camera was equipped with a good-quality Varamat 9-27 mm Fl.8 zoom lens, which magnified the image about four times. It was loaded with 16 mm Kodachrome II Safety film, an outdoor film that he had previously used to take movies of his grandchildren playing. The camera exposed only one half of the film at a time, at which point it was reversed. The camera lab would cut the film down the middle, returning two reels of 8 mm film from the original 16 mm reel.

Zapruder began searching for a spot to stand near the north grassy knoll area. He would later tell the commission, "I tried one place and it was on a narrow ledge and I couldn't balance myself very much. I tried another place and that had some obstruction of signs or whatever was there and finally I found a place further down near the underpass that was a square of concrete."[4] After considerable effort, Zapruder found a location on the concrete steps of the pergola. He tested the camera by shooting a few frames of his receptionist, Marilyn Sitzman. Sitzman suggested that her boss should try standing on the small concrete abutment forming part of the pergola built on the slope of the north hill of the plaza, halfway between the Texas School Book Depository building and the railway. But Zapruder hesitated. He suffered from vertigo and feared he could not maintain his balance. Sitzman offered to hold onto his coat to steady him. It was from this position, about 65 feet from the center of Elm Street, that Abraham Zapruder captured the most famous 8 mm film in history.

At 12:30 p.m. John F. Kennedy's motorcade arrived from the east from Houston Street, and turned left onto Elm Street toward the center of the open park area. Zapruder began to film the motorcade, then realized that he was only recording motorcycles in the lead and not the presidential limousine. He stopped and began filming again, never taking his eye and viewfinder off the presidential limousine until it passed under the Triple Underpass, where the rail line ran above, and continued on to Parkland Hospital.

Zapruder was filming with a telephoto lens that magnified everything he saw. This film today still remains the best photographic record of the assassination. The images are in color, are clear, and depict almost every crucial event that transpired in Kennedy's motorcade. Kennedy can be seen reacting to almost every shot that struck him.

After witnessing the most horrific sight of his life, Zapruder retreated, unimpeded, back to his office in the Dal-Tex Building. No one tried to stop him. No one asked for his camera. No one noticed him. His film and camera were not confiscated. He now had in his possession a treasure of national significance.

He knew then that he was sitting on a goldmine.

At that time, no one knew about the film except Zapruder and his receptionist, Marilyn Sitzman, and another employee, Beatrice Hester. Shortly after the assassination, however, Sitzman and Hester were approached by a reporter in Dealey Plaza, Darwin Payne, who worked for the *Dallas Times Herald*. Sitzman told Payne about Zapruder and the existence of the film of Kennedy's shooting. Payne proceeded to Zapruder's office, where he took hasty notes before phoning his newspaper with the story of the decade. A second reporter, Harry McCormick of the *Dallas Morning News*, also interviewed him. Within an hour, Zapruder was appearing on WFAA, the ABC television affiliate. He had arrived there in hopes that the station could develop his film. After his live television interview, Zapruder took his film to the Kodak plant for development. He had a long day ahead of him, and ultimately he would visit seven different offices and plants before obtaining an original print and three copies of his amateur film.[5]

Sitzman herself would later give a description of what she saw as she steadied her boss that day in Dealey Plaza. Recalling her observations to Josiah Thompson in 1966:

> There was nothing unusual until the first sound which I thought was a firecracker, maybe because of the reaction of Pres. Kennedy. He put his hands up to guard his face....And the next thing that I remembered clearly was the shot that hit directly in front of us, or almost directly in front of us, that hit him on the side of his face....Above the ear and to the front....And, we could see his brains come, you know, his head opening.[6]

Sitzman's recollection matches the images depicted on Zapruder's film exactly, despite the fact that the film was not shown to the public until 1975. Her description of the location of the shot was both precise and accurate. Note also her recollection, seen from a direct vantage point, that the shot struck the side of Kennedy's face. Frame Z313 shows this clearly.

After talking with Zapruder, Harry McCormick proceeded to the sheriff's office to cover developments there. It was here that he encountered Secret Service agent Forrest Sorrels, whom he had known "for many years."[7] McCormick told Sorrels about the film and the two left for Zapruder's office, arriving a little after 1:00 p.m. By this time, President Kennedy had already been pronounced dead, but the world had yet to be informed of this momentous redirection of our nation's history.

Zapruder's partner, Erwin Schwartz, had already joined him at his office. With him were two Dallas policemen, Darwin Payne, receptionist Lillian Rogers, as well as other news media personnel. Sorrels entered Zapruder's office alone. Zapruder was extremely distraught but still had his wits about him. Sorrels asked for a copy of the film and Zapruder agreed under the condition that it would be used only by the Secret Service and would not be shown to any newspapers or magazines. Zapruder made it clear that he expected to sell the film for whatever he could get for it. While they were talking, Zapruder's camera and film were securely locked in a safe in his receptionist's office.[8]

It was at this point that McCormick, who was trying desperately to purchase the film for his company, suggested that his newspaper, the *Dallas Morning News*, could develop the film. The two policemen took Zapruder, Schwartz, Sorrels, and McCormick to the news offices. No one there wanted to try to develop the film. The newspaper's television station, WFAA-TV, which was next door, could only process 16 mm film. After Zapruder's interview at the station, the six men piled back into the squad car and headed over to the Eastman Kodak Processing Laboratory at Love Field. They arrived in time to see Air Force One, bearing Kennedy's body, depart for Washington, DC.

Richard Blair, a technician from the service department, took Zapruder into a darkroom and removed the exposed film from his camera. Blair then handed the spool to Kathryn Kirby, who inserted the film into a processing machine that had been dedicated to processing only this film. Secret Service agent Sorrels remained in the room with the Kodak technicians as they processed the film.[9]

While they were waiting for the roll of film to dry from the processing chemicals, Sorrels called his office and discovered that a man had been arrested for the murder of a police officer in Dallas and that he was needed back at the station. Before Sorrels left, he told Zapruder, "If it comes out, get me a copy."[10]

The film was processed and readied for review within an hour. Phillip Chamberlain, the laboratories production supervisor, and Zapruder reviewed the film with a faster-than-normal speed projector. During the quality control review, Zapruder asked Chamberlain if he could have three copies made of the film. The Kodak plant did not have the capability to make copies, so a call was made to the Jamieson Film Company in Dallas. Jamieson agreed to make the copies provided they could use 16 mm film. After further review of the film, Zapruder asked to sign an affidavit attesting to the validity of the work performed at the Kodak lab. Chamberlain's affidavit stated his lab had processed Zapruder's film and that it was "not cut, mutilated or altered in any way during processing."[11]

After a brief stop at his office, Zapruder and Schwartz drove to the Jamieson Film Company, arriving at about 6:30 p.m., their precious film in hand. Jamieson and his staff proceeded to produce three copies of the film with Zapruder watching every step. Because the optimum conditions for processing of Kodachrome film were not familiar to them, the company's staff used different processing settings for each copy. Copy 2 was determined to be the best; however, none of the copies retained the images around the sprocket holes. The processing was finished shortly after 7:30 p.m. Again, Zapruder requested that the laboratory manager swear out an affidavit as to the authenticity of the copies. The affidavit, signed by manager Frank Sloan, stated that exactly three copies were made.[12] No one at Jamieson viewed the film or its copies.

Schwartz and Zapruder immediately drove back to the Kodak laboratory to have the duplicates developed. The original 8 mm film had already been tagged with number 0183, and the 16 mm copies were assigned numbers 0185, 0186, and 0187 to the end of their filmstrips. Once more, Zapruder requested signed affidavits from the staff. The processing was completed around 9:00 p.m. Zapruder then had to travel to the Secret Service office in downtown Dallas to drop off two copies of the film.

When he arrived home, sometime after 10:00 p.m., Zapruder showed the original film on his 8 mm projector to his family. By 11:00 p.m., he had received a call from Richard Stolley of *Life* magazine, who was seeking to acquire the historic film. Stolley recalled:

> I picked up the Dallas phone book and ran my finger down the Zs, Bingo! There it was—"Zapruder, Abraham." I called the number. No answer. I called that number every fifteen minutes for the next five hours. Finally, at about eleven, this weary voice answered. I identified myself.... I said, "Is it true that you photographed the assassination?" "Yes." I said, "Can I come out and see it?" And he said, "No."... So I said, "Fine. When can I see it tomorrow?" He said, "Come to my office at nine."[13]

The next morning, he permitted the film to be shown to his employees coming on shift. Two of the copies, numbers 1 and 3, Zapruder had given to the Secret Service the night before. One of these copies (3) was sent to Washington, DC, later that night while the other copy remained with the Dallas Secret Service. The FBI had obtained its copy of the film from the Secret Service. Zapruder had retained the master and one copy.

By 8:00 a.m., November 23, Zapruder had set up to show the film to a group of Secret Service agents in his office. Richard Stolley arrived at this time, an hour before Zapruder had told him he could come, and Zapruder invited him to watch the film with them. After viewing the film, Stolley was determined to acquire it for *Life* magazine. He recalled the atmosphere at Zapruder's company that morning:

While we were in there talking to Zapruder, other reporters were starting to come into the place.... So then we all gathered out in the hall, and people said, "We want to talk to you, Mr. Zapruder, about this film." And he said, "Mr. Stolley here was the first one to contact me, so I'm going to talk to him first." And the others went berserk, just began screeching, shouting at him, said, "Promise you won't sign anything Mr. Zapruder."...We went into his office and I thought, "There's no way in the world I'm going to walk out of this office without that film in my hand." [Zapruder] was a very genial man. He's been born in Russia and he came and he worked in the garment industry.... So we talked for a while, then I would throw in a new figure. He made it very clear that he was doing this [because] the garment business was touch and go, so he wanted to secure the financial future of his family insomuch as he could.... I was authorized to go to $50,000, which was a hell of a lot of money back then.... And just about the time I reached $50,000, somebody kicked the door.... I said, "Mr. Zapruder, I'm going to be honest with you, this is as high as I'm authorized to go.... And then "Bang!" And he looked at me and said, "Let's do it."[14]

Stolley walked over to the typewriter in Zapruder's office and typed out a six-line contract. Both men signed it. Stolley then left through a back door, leaving Zapruder to face an angry crowd of reporters in the hall. Richard Stolley of *Life* magazine had purchased the original print of the film from Zapruder for $50,000. He immediately had someone from the Dallas bureau courier it to the Chicago printing plant by commercial plane. Still black-and-white photographs taken from the film were published in *Life* magazine on November 25, the Monday morning following the assassination. Frame Z313, the grisly frame showing the head shot, was omitted out of respect for the Kennedy family.

Over the years, a body of literature has arisen that claims the original Zapruder film was altered. Several books have been devoted to this subject together with a number of articles, conference papers, and Internet sites. An excellent discussion of this literature is provided in University of Wisconsin professor of history David R. Wrone's thorough and sound historical treatment, *The Zapruder Film: Reframing Kennedy's Assassination.*[15]

In his book *Best Evidence*, published in 1980, David Lifton alluded to several instances of film alteration including missing frames, frames out of order, and splicings. In 1984, Philip Melanson published an article in the magazine *Third Decade* titled "Hidden Exposure: Cover-Up and Intrigue in the CIA: Secret Possession of the Zapruder Film," which made similar claims as Lifton.[16] By 1997, a collection of alterationist essays had appeared in *Assassination Science: Experts Speak Out in the Death of JFK*, edited by James H. Fetzer.[17] This compendium contains an article by medical doctor David Mantik, "Special Effect in the Zapruder Film: How the Film of the Century Was Edited." These works have been widely criticized, however, as documents that distort facts and suffer from serious errors and omissions. A most telling argument against these claims has been advanced by Harold Weisberg in his books *Whitewash* and *Never Again.*[18] It makes no sense for a government that has repeatedly sought to prove that Kennedy was killed by a single assassin to alter a film in such a way as to demonstrate that shots were fired from additional locations in Dealey Plaza, which analysis of the Zapruder film demonstrates in multiple ways.

Wrone argues that, to begin with, contrary to the claims of the alterationists, there was no opportunity to steal the film.[19] Neither was there time or technology available to alter it. Further, the suspected agencies, the FBI, the CIA, and the Secret Service, initially showed little or no interest in the Zapruder film. The Secret Service agent Forrest Sorrels, who had accompanied Zapruder to the Kodak film-developing facility, had left before the film was developed. Zapruder delivered it himself to their Dallas office. The FBI didn't even request the original film, later borrowing the Secret Service copy to view it. The CIA did not even begin to study the film until after photographs taken from it had appeared in *Life* magazine.

No opportunity existed in the film's chain of custody to enable conspirators to filch and alter the film. Abraham Zapruder and his partner, Erwin Schwartz, controlled the original film from the time Zapruder recorded it on November 22 until he sold it the next morning in his office to Time, Inc., which owned *Life* magazine. Zapruder, Schwartz, and Sorrels watched the Kodak technicians as they processed the film. After development, Zapruder required the film processors to validate their work with notarized statements. By 9:00 p.m., Zapruder had the processed film plus three copies to take home with him. By 11:00 p.m., he had the film and one copy in his possession in his home, finally going to bed around midnight. He arose by 7:00 a.m. and had the film in his office by 8:00 a.m. the next morning. Conspirators would have had to have broken into Zapruder's home while he slept, stolen the film, flown it to Washington, examined the film, altered the appropriate frames, re-photographed it and reprocessed it to produce a new original, then flown it back to Dallas, broken into his home again, and replaced the film all within a six- to seven-hour period. Since the flight from Washington to Dallas was three hours long, this task would have been virtually impossible.[20]

It is very unlikely that Time, Inc., would have subsequently done anything to jeopardize the flow of income it expected from the film. Allowing someone to alter it would have destroyed their investment and ruined their revenue stream. The claims of alteration of the film lack both hard facts and plausibility.

In 1996, Roland Zavada, an expert in film production and a former employee of Kodak, began a thorough technical study of the authenticity of the Zapruder film at the behest of the Assassination Records Review Board (ARRB), which was established in October 1994 and commissioned to release all previously classified documents related to the Kennedy assassination.[21] As a product engineer for Kodak, Zavada had worked on the development of the very film stock that Zapruder had used in his camera, Kodachrome II.

Kodachrome II was specifically designed for outdoor exposures. It is a reversal film, which means that it produces a positive image upon development as opposed to a negative. It can be viewed therefore immediately after processing. To compensate for the brightness levels of a film projection in a darkened room, the film is designed to enhance the contrast levels of the original images beyond what the light levels were of the scene in the viewfinder. This contrast enhancement differs depending on the color of the original image. Red colors would exhibit the highest contrast densities, while greens and blues would exhibit a lower density on the film. If someone tried to copy one Kodachrome II film onto another, this contrast would get boosted twice. Moreover, the contrasts of the colors would get boosted a second time by different amounts, causing the image to look distorted both in overall contrast and in color balance. Zavada was able to determine that the Zapruder film was from original Kodachrome II film stock. If someone had tried to make alterations and "finish" those alterations onto Kodachrome II, the images would have appeared noticeably and badly distorted. This was not the case for the Zapruder film.[22]

Special-effects wizardry in Hollywood is typically performed on large, 35 mm or greater film formats, but not on 8 mm amateur films. The 8 mm film would have been too small for the matte artists to work with. Blowing the film up to 35 mm and reducing it back down to original Kodachrome II 8 mm stock would have produced distinctive contrast problems. Any attempt to make alterations in this manner would have resulted in telltale structural characteristics that would be apparent to a trained observer, if not an amateur viewer.

In a 2003 letter responding to allegations of film alteration made after release of his 1998 report, Roland Zavada commented on the possibilities that the original Zapruder film could have been altered.[23] Hollywood special-effects artists use a "family of films" specifically designed to allow them to achieve their results. These films work together to produce a final film with the right lighting and contrast. A great deal of effort has been invested to find the right emulsions to process the work from step to step, from film to film in the process. A series of films is employed, so that a special-effects production uses film stocks in a cycle: camera original, interpositive, internegative, and print film. For anyone to establish a believable case of forgery, a series of stocks must be found that were available in 1963 and that could process a Kodachrome II film through this alteration cycle and end with a Kodachrome II film where the colors, lighting, and contrast looked natural. Kodachrome II film was never designed with this alteration process in mind.

The original Kodachrome reversal film was designed to be projected by a hot light source, not as a print medium. Because it produced a positive image, as opposed to a negative, it was never intended to be used as an intermediate film in a production series. Zavada stated:

> The film was not designed for printing response so that its dye set matched the spectral sensitivity of laboratory intermediate negative or positive films. A reversal duplicating film was available, but that was for direct simple copies, and not expected to be used as an intermediate. Further the film's daylight sensitivity; contrast and spectral characteristics do not render it receptive for use as a "print" medium—hence, one "hell-of-a-problem" for someone trying to replicate a Kodachrome original (Note: the goal now being to create a "Kodachrome original") by using special optical effects![24]

It would be extremely difficult for someone to alter the film using a film-processing technology not known to exist in such a way as to show no signs of alteration under microscopic analysis. Alterations to the film's fine grain structure as well as to image modulation and contrast would be apparent through microscopic examination. Further, the film used by Zapruder would have been difficult to "register," or align with the sprockets after alteration. A reference perforation or edge would have been required, neither of which were available on Zapruder's film. Therefore, the tight tolerances required for undetectable film alteration could not be achieved for this type of film.

Zavada was not specifically asked by the Assassination Records Review Board to determine if the film had been altered. The goal of that work was not to produce a "statement of authenticity," but was instead to analyze the evidence. The goal of the study was to create a knowledge and fact database from which other researchers could draw their own conclusions. Zavada's work centered on addressing the vintage of the films, the processing technology and markings, the printing technology and characteristics, and the camera image capture characteristics. However, the probability of alteration by applying laboratory optical effects or simple printing techniques (to remove selected frames) after transfer of the original to an intermediate, as proposed by some researchers, was reviewed. Zavada's careful viewing of multiple scenes and his knowledge of optical effects technology convinced him that a dissertation on the probability of alteration was not needed. Zavada was convinced that the film designated in the archives as *Zapruder-in-camera-original* was exactly that.

After Zavada's work under contract had been completed, he was free to express a personal opinion as to the authenticity of the film. He stated:

> The film that exists at NARA [National Archives and Records Administration] was received from Time/Life, has all the characteristics of an original film per my report. The film medium, manufacturing markings, processing identification, camera gate image characteristics, dye structure, full scale tonal range, support type, perforations and their quality, keeping shrinkage and fluting characteristics, feel, surface profile of the dye surface. It has NO evidence of optical effects or matte work including granularity, edge effects or fringing, contrast buildup etc.

Simply stated:

> There is no detectable evidence of manipulation or image alteration on the *"Zapruder in-camera* original"* and all supporting evidence precludes any forgery thereto.[25]

Roland Zavada's prodigious technical expertise and detailed study of the Zapruder film is a powerful testament to its authenticity. Historian Michael Kurtz writes in his book *The JFK Assassination Debates*:

> Although I have great respect for the work of such forgery advocates as David Mantik, I must agree with the proponents of the Zapruder

film's authenticity. I am persuaded first and foremost by the work of Roland Zavada, who has destroyed every argument made by the forgery theorists. His painstakingly detailed report confirms that the film contains no evidence of alteration, nor does it depict photographic anomalies that cannot be explained.... Finally, David Wrone's lengthy and exhaustive study of the history of the Zapruder film argues persuasively for its authenticity.[26]

In science, everything is open to question all the time. The existence of the film allows present and future generations of scientists and researchers to examine the record for signs of alteration using the latest and greatest technologies. To date, no technology or expert analysis has revealed the Zapruder film to be anything but an original photographic record of events that transpired in 1963. As such, this film record is suitable for scientific analysis, and together with the 1963 Dictabelt recording, constitutes the most complete extant visual and aural record of the assassination of John F. Kennedy.

CHAPTER 9

THE SECOND RIFLE IN DEALEY PLAZA

The American Dream has run out of gas. The car has stopped. It no longer supplies the world with its images, its dreams, its fantasies. No more. It's over. It supplies the world with its nightmares now: the Kennedy assassination, Watergate, Vietnam.

—J. G. Ballard

The ideal form of government is democracy tempered with assassination.

—Voltaire

Nnovember 22, 1963. The Friday before Thanksgiving and all across the country Americans awoke in warm anticipation of the upcoming holiday. Since the Cuban Missile Crisis had been resolved a year before, the country had experienced a relative state of peace. Few Americans had even heard of Vietnam, and the phrases "terrorist" and "suicide bombing" were not yet in common use.

It was an era of innocence. Democracy and good had prevailed in Europe and Japan just fifteen short years before. The blandness and inertia of the Eisenhower years had given way to the new frontier of a young and charismatic new president, John Fitzgerald Kennedy, who only recently had challenged the country to "send a man to the moon and return him safely to earth before the decade is out."

But that innocence was shattered forever when shots rang out in Dealey Plaza in Dallas, Texas, shortly after 12:30 p.m. In a matter of moments, the dreams of a nation turned into a nightmare that would affect an era and would ripple around the world. The assassination of JFK would become one of those rare benchmarks in time by which all else is measured. Yet, nearly fifty years later there are more questions than answers, and myths and legends compete with reality in the search for truth.

Who killed Kennedy and why? As discussed at length in chapter 2, the conclusions of the Warren Commission are untenable. Failure of the single-bullet theory implies a second gunman. But before anyone can attempt to find the truth of the assassination, it is first necessary to determine the source of the final, fatal shot. If the second murder weapon and the true locations of the assassins can be identified, there will be a fighting chance to uncover the truth. The postulation is this, simply: the rifle alleged to have fired the final, fatal shot was not capable of doing so. Careful physical analysis of rifles and ammunition available in 1963 suggest there was only one rifle that could have fired the fatal head shot that killed Kennedy.

This chapter does not purport to solve the mystery; rather, it is to develop a scientifically solid baseline from which to work forward. But finding the truth behind the assassination is a difficult proposition because the confusion and the controversy surrounding Kennedy's death has been overwhelming, a veritable Twilight Zone of conspiracy theories, magic bullets, missing witnesses, conflicting testimonies, faked x-rays, and contradictory evidence. To navigate this murky landscape a guide is needed. That guide is the law of nature.

The laws of physics don't change. They are the same at all times and in all places. The same physical laws that apply here on Earth pertain throughout the universe. Wherever and whenever astronomers look to the heavens, they see the same forces, the same laws of motion, and the same natural phenomena that are observed here on Earth. The gravity felt and experienced every day is the same force that keeps the planets in their orbits around the sun, the sun in its orbital course throughout the galaxy, and the galaxy in revolution about its neighbors. The spectra of elements observed in our laboratories are the same as those observed in distant stars and nebulae. The scale of the universe is immensely larger, yet the rules governing its operation are precisely those that rule our mundane earthly existence.

When telescopes scan the stars and vastly distant objects like the Andromeda galaxy, for instance, they literally look back in time. The light seen today left Andromeda over two million years ago, at a time when the first men were making the first crude pebble tools at Olduvai Gorge. When astronomers observe a quasar like 3C 273, they look back more than two billion years to an era when the earth was barren, the air unbreathable, and the seas were seething with only single-celled algae and bacteria. With the orbiting telescope of the Hubble Deep Field survey, galaxies are seen at a time shortly after the beginning of the universe, 13.7 billion years ago. Its images exhibit a young, dynamic, energetic universe, the first primeval galaxies, bursting supernovas, and gamma-ray-producing black holes, but it is clear that the laws that governed that universe are precisely the laws that regulate ours today.

At whatever time or place scientists gaze into the cosmos, whether they observe nearby stars, distant supernovas, or fireflies in their backyards, they always see the same fundamental forces, the same laws of motion, and the same physical constants. Although the universe is in eternal flux, continuously evolving, the principles that govern its behavior remain the same. It is this observed eternal immutability of the timeless universal forces that allows us to conclude with confidence that the laws of physics are no different today than they were moments after the big bang, the day Isaac Newton was born, or on November 22, 1963.

Science solves many cases. Countless criminals have been brought to justice and innocent men exonerated by the science of DNA analysis. Unfortunately, the science of DNA can't help here. But other branches of science can. To attempt to resolve the Kennedy assassination using the Zapruder film, the branches of physics called dynamics and kinematics are required. These branches characterize forces and motion using precise mathematical expressions.

Modern science is practiced and expressed using mathematical formulae. Mathematics is the language of nature. In order to resolve the truth of the assassination, one must apply physical laws that are best expressed as algebraic formulas. If you find math, especially algebra, about as abstruse as Sanskrit, please bear with me. It will be easier than you think and it will lead us to the right answers. This is the only way to discover what really happened to Kennedy.

What are the laws of physics that apply to the Kennedy assassination? Because Kennedy's head recoils backward at the moment of impact of the final shot, as shown in the Zapruder film, and because the bullet has both mass (weight) and velocity (speed) prior to impact, the applicable physical principle is known as the conservation of momentum. This backward motion of the head is also captured on the Nix and Muchmore films, as well as on the Mary Moorman Polaroid photo, taken within moments of the final fatal shot.

Momentum is a vector. Anything moving has a momentum vector associated with it. The momentum vector characterizes the speed, mass, and direction of movement. A tractor trailer moving along the highway at 60 miles per hour will have a substantial momentum vector associated with it because it is so massive. A car moving toward it from the other side of the highway will have a much smaller momentum headed in the opposite direction, even though the car may be moving at the same speed as the truck. If the two collide, the disparity in the two momentum vectors will become painfully apparent to the driver of the car.

Vector Addition

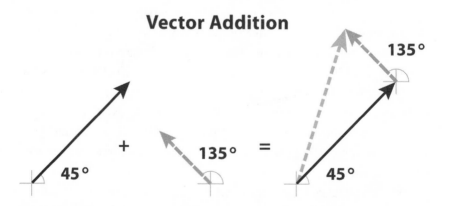

Figure 18: The dashed line shows the resultant velocity vector when two initial velocity vectors are summed. Courtesy of Linda Alvey-Chambers.

THE SECOND RIFLE IN DEALEY PLAZA

Expressed mathematically, the momentum of an object, a person, or a projectile has a magnitude (mass times velocity) and a direction. A vector in physics is represented as an arrow in space. Its length is proportional to its magnitude, while its orientation specifies its direction of motion. Vectors allow for the expression of physical quantities that have directional information in a convenient graphical format. Other mathematical formulations are possible, but vectors are far simpler to work with than other expressions.

In any isolated physical system, including a self-interacting one, momentum is always conserved. Unless a force operates to change an object's speed or direction, its momentum must remain the same. If it collides with another stationary object, that momentum must be apportioned between the two objects. Stated precisely, the final momentum of all the components of an interacting system must equal the initial total momentum of the system's elements, provided no external forces are acting. The momentum of interacting objects is combined by placing the vector arrows head to tail and drawing a new arrow from the initial tail to the final head. The magnitude of the momentum vector is always given by the mass of the object (m) multiplied by the velocity (v), denoted as mv (in physics, two variables written next to each other denotes multiplication of the quantities). Momentum conservation is a cornerstone of modern physics and has been known since the time of Galileo.

A car is traveling north on I-95 from Washington to Boston. It has momentum in a northerly direction. The magnitude of that momentum is determined by multiplying the mass (weight) of the car, say, 3,000 lbs., times its velocity of 55 miles/hour, yielding 165,000 lbs.-miles/hour. This seems like a large number and it is. If the car were to crash into another car head-on with this momentum, the two vehicles would be totaled beyond repair. If the car changes direction, its momentum will also change. This is known as acceleration, which accounts not only for changes in velocity but for changes in direction as well. If the car slows down or speeds up, its momentum also changes due to acceleration.

In the case of a rifle shot, the bullet is moving prior to impact and therefore possesses kinetic energy as well as momentum. The analysis consequently requires another conservation law for energy. Energy itself can take many forms. In can be kinetic (associated with motion), potential (associated with height), electrical (as produced by a battery), thermal, nuclear, or chemical. It can transfer from one object to another (as in hitting the eight ball with your cue); it can manifest itself in myriad and diverse ways, but it is always conserved; that is, energy can be neither created nor destroyed, it just changes its form (although matter can be converted into energy and vice versa in nuclear processes, according to Einstein's theory of relativity).

One of the best illustrations of energy conservation was given by the late Nobel laureate Richard Feynman. In his famous *Lectures on Physics, Volume I,* Feynman uses the analogy of Dennis the Menace's blocks to represent energy and its different forms. Dennis the Menace scatters his play blocks all over the house, under the bed, beneath his blankets, in the driveway, on the floor, in the bathtub, and so on. The blocks, like energy, cannot be destroyed. Sometimes they are apparent, as when you trip over them as you walk across the floor. This is like kinetic energy, the form of energy associated with motion, and is obvious, especially when an object with lots of kinetic energy, like Barry Bonds's baseball bat, contacts another object, like Nolan Ryan's fastball, and sends it to the parking lot. At other times energy is subtly hidden, invisible, yet it is there nonetheless like the potential energy associated with the height of an object (if you drop the object, it will accelerate until it hits the ground); think of the blocks hiding under the blankets. Sometimes the presence of energy is not obvious, as when the blocks are under the bathtub water. The blocks cause the level of the bathwater to rise, so that it is above the ring of dirt in the tub that you've been meaning to scrub off. The blocks, like energy, reveal themselves if you probe carefully, as when the energy of friction manifests itself as heat. It can't be detected until you touch the surface with your hand. Energy, like blocks, must always be conserved but can manifest itself in diverse and varied forms. Like the Cheshire cat, sometimes you see energy, sometimes you don't.

Taken together, momentum conservation and energy conservation form a powerful duo that allows physicists to solve a variety of challenging and useful problems. A well-known example is the rocket equations. Momentum conservation is the key to rocket technology because the momentum of the gases exiting the rocket must equal the forward motion of the rocket plus its fuel and payload. The faster the reaction product gases exit the tail, the faster the rocket moves up through the atmosphere. As it burns its fuel and climbs higher and higher, the rocket converts its kinetic energy into potential energy in Earth's gravitational field. If its engines were to fail, it would fall back to Earth, accelerating to higher and higher velocities until it impacted the surface. In this scenario, its potential energy would be converted back into kinetic energy. If, on the other hand, its engines continue to accelerate the rocket and it can achieve a sufficiently high rate of speed, it can escape Earth's gravity entirely. The velocity needed for any object to escape from Earth is found using the law of energy conservation and works out to be about 11.2 kilometers/second (6.95 miles/sec). Momentum conservation is then used to determine how fast the exhaust gases must exit, given their mass, to achieve escape or orbit for a given payload.

This is rocket science. Without these laws, satellites and the rockets that launch them would be a hit-or-miss proposition. It would be impossible to place satellites into orbit so that they would remain there. Today, much of our technology depends on the presence of these satellites, which are crucial for communication, weather monitoring, and the functioning of GPS (Global Positioning Satellite) technology. The next time you find yourself listening to satellite radio in your car or find your way on a dark road on a cold rainy night using GPS navigation, think of conservation of energy and momentum. It is our deep and profound understanding of these laws that makes our global electronic civilization possible.

Consider the Kennedy assassination through the revealing lens of the conservation laws. At the time of the final, fatal shot, Kennedy was slumping forward in his seat, his head inclined at about a 20-degree angle, as shown in the Zapruder film, frame Z312 (see figure 17). Dr. Art Hoffman, an expert quoted in Vincent Bugliosi's book *Reclaiming His-*

tory, argues that a bullet striking Kennedy from the front must push against gravity to drive his head backward, a scenario equivalent to Kennedy lying prone on the ground, the bullet striking him from below, raising him against the full force of gravity.[1]

But Kennedy was not supine at the moment of impact. Instead, his head was inclined slightly forward, suspended on a pivot—his spine and neck. His head was supported against gravity by his body, so it wasn't necessary for the bullet to push against the full force of the weight of his head and body, only against the small component of gravity due to the angle at which Kennedy's head was listing.

The situation is analogous to pushing a child on a swing. It's easy to push the child from a vertical position, but it's very hard to lift the child in the swing by pulling on the supporting ropes. Pushing requires little force because the child's weight is supported by the ropes. You need only overcome the inertia due to her mass, not her weight due to Earth's gravitational pull. Of course, the easiest time to push the swing is when it is at its highest point, when the potential energy is at a maximum and the kinetic energy at a minimum, just at those points when it starts swinging forward or back.

The physics behind the child's swing is similar to that of the ballistic pendulum, a simple device for measuring the velocity of a bullet in the laboratory. The ballistic pendulum consists of a block of wood suspended by a wire. When a bullet strikes the block and remains inside it, it causes the pendulum to swing upward, converting kinetic into potential energy in raising the block against gravity. This is due to conservation of momentum. The momentum of the bullet is transferred to the block. Because the block is raised by the incoming momentum of the bullet, as the block slows and reaches a stop at a higher point, its imparted kinetic energy is converted into potential energy in the form of its height. Therefore, the height the block reaches after the collision gives a measure of the kinetic energy of the block and bullet immediately after the impact (see figure 19). The momentum of the incoming bullet is automatically determined since it must match that of the recoiling wood block with the bullet trapped inside, due to the prin-

The Ballistic Pendulum

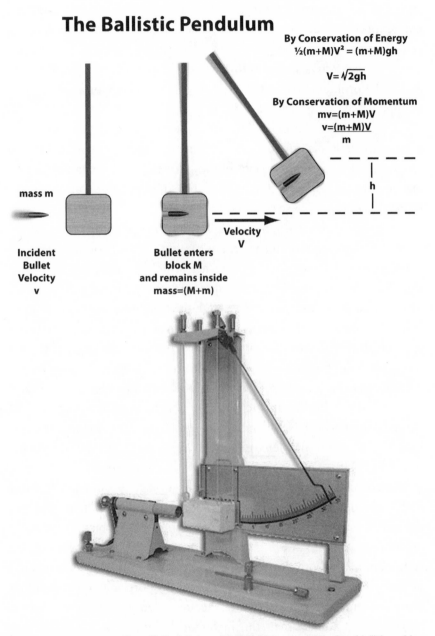

By Conservation of Energy
$$\tfrac{1}{2}(m+M)V^2 = (m+M)gh$$

$$V = \sqrt{2gh}$$

By Conservation of Momentum
$$mv = (m+M)V$$
$$v = \frac{(m+M)V}{m}$$

mass m

Incident Bullet Velocity v

Velocity V

Bullet enters block M and remains inside mass=(M+m)

h

Figure 19: The ballistic pendulum. (Top) Schematic of the action of ballistic pendulum.
Courtesy Linda Alvey-Chambers.
(Bottom) Photograph of a working ballistic pendulum device.

ciple of momentum conservation. Then, dividing the bullet's momentum by the mass of the bullet determines the projectile's velocity from the gun.

A similar analysis applies to the assassination. The only difference is that the problem is inverted; Kennedy's head is supported on a pivot as opposed to being suspended from a wire.

The complication of the assassination, however, is that the initial mass of the bullet is unknown. The caliber of the gun that fired it is also unknown. However, reasonable initial guesses for these values can be made using the Zapruder film. Then, applying the conservation laws, it is possible to test if these starting assumptions prove out. This is a legitimate way to proceed when there is no reliable starting point.

If you have ever tried shooting a can with a rifle, you have noticed that the can did not appear to move when struck by the bullet. When you examined the can, you found it had clean holes punched in both sides. Momentum was conserved even though the can didn't move because the bullet carried away its initial momentum when it exited, suffering no appreciable loss in its velocity as a result of its passage. The total momentum of the stationary can plus the moving bullet before the impact was the same as the can and the bullet system after the impact.

However, as in the ballistic pendulum, if the impacted object moves appreciably it is because the impacting projectile has remained trapped inside it. When this happens, the projectile goes from its initial velocity, upward of 2,000 feet per second for a rifle bullet, to effectively zero. However, since momentum must be conserved, the impacted object with the bullet now trapped inside must recoil in the initial direction of the bullet, but with a much lower velocity since the object is typically far more massive than the bullet. Because momentum consists of the compounding of both the mass and the velocity of the object, when the mass increases the velocity must decrease proportionately to maintain the same momentum. Therefore, the impacted object recoils with a much lower velocity than the incoming bullet, in a manner that is clearly visible to the unaided eye in contrast to the invisibility of the high-velocity bullet prior to impact.

THE SECOND RIFLE IN DEALEY PLAZA

In the Zapruder film, Kennedy's head clearly and distinctly recoils backward and sideways, at about a 45-degree angle with respect to the axis of the vehicle, due to the impact of the final, fatal round as depicted in frames Z313 to Z323 of the film shown in figure 20. Because Kennedy's head recoils backward at the moment of impact, it is reasonable to conclude, based on the law of conservation of momentum, that the bullet that struck him arrived from the front side of his head, remained trapped inside, and never exited.

This is consistent with the evidence from the films. No trace of a bullet exit wound is apparent on either the Zapruder film or the Mary Moorman Polaroid photograph that gives a clear view of the rear left side of the president's head moments after the final bullet impact. A rifle round tumbling inside Kennedy's head would do substantial and visible damage upon exit, yet no damage can be observed anywhere on the back or left side of his head. Neither Mrs. Kennedy nor anyone else was struck by bullet fragments exiting Kennedy, consistent with the film records. The doctors at Parkland Hospital noted no wounds of any kind on Kennedy's face, the rear of his head, or left side of his head. The lack of any exit wound is consistent with the conclusion that the round that killed Kennedy entered above and in front of his right ear, but never left his head. This result is further shown to be consistent with the medical evidence, indicating that copious metallic fragments were found in Kennedy's brain.

Large-caliber bullets, like those of the .26 Mannlicher-Carcano, have considerable mass and penetrating force while small-caliber rounds, like the .223, are substantially less massive and lack the punching power of their larger-caliber cousins. Small rounds, however, tend to remain inside the body, doing extensive internal damage. This is the principle behind the .22-caliber M-16. Smaller objects have a larger surface area to volume ratio for a given shape. Since a bullet slows in proportion to its surface area through friction, small rounds come to rest inside the body much more readily than large-caliber bullets. If the round is frangible, designed to come apart on impact, it will fragment when it hits a thick bone like the skull, driving lead particu-

Figure 20: Z313 (top) and Z323 (bottom) show the backward movement of
Kennedy's head in response to the impact of a high-speed rifle round.
Zapruder film © 1967 (renewed 1995). The Sixth Floor Museum at Dealey Plaza.

lates and skull fragments into the brain, where they will inflict massive carnage. A standard frangible round design consists of powdered lead housed in a copper jacket (what we see on the outside of the bullet). Based on the head recoil seen on the film records and the eyewitness evidence of substantial metal fragments in Kennedy's brain, the bullet that killed Kennedy was likely a small-caliber frangible round that fragmented inside his head.

Frame Z313 of the Zapruder film clearly shows multiple jets of blood, bone, and brain matter exiting from the entry wound just above Kennedy's right ear. Pronounced tissue ejecta are characteristic of the impact of a high-velocity round, which carries a great deal of kinetic energy. Because kinetic energy is proportional to the square of a projectile's velocity, a bullet traveling at 4,000 ft/sec will possess four times the kinetic energy of a similar round traveling at medium velocity, say, 2,000 ft/sec, like the Mannlicher-Carcano. The bullet that caused the extensive tissue jetting observed in frame Z313 was a high-velocity round traveling at or near 4,000 ft/sec. This type of conspicuous damage is a distinguishing feature of a high-velocity round, as is similar to that observed when a high-velocity bullet hits a watermelon.

Once the bullet penetrated Kennedy's skull, it pressed hard against the internal liquid blood and brain tissue, causing the pressure to spike within it. It was this dramatic pressure rise inside his head that caused the jetting of blood and tissue observed in frame Z313. Tissue, blood, and brain matter extruded back through the entrance hole created by the bullet, driven by both internal systolic pressure normally in the brain cavity and induced pressure from the bullet impact. The brain matter and blood, being ductile, fluid, and under extreme pressure, found an exit, like water squirting from a hose. The entrance hole of the bullet was the only available outlet for the pressurized fluid in Kennedy's head. With his head recoiling backward, ejecta from his wound also traveled behind him and to the sides and impacted people around him.

But what weapon actually fired this fatal shot? A prime candidate for the high-speed rifle with high accuracy and a small-caliber round is the .220 Swift, a favorite assassination weapon of the 1960s.[2] The rifle was originally produced in 1935 as a superspeed .22 centerfire by Winchester. It is extremely accurate and is used today for hunting small game and deer. Thirty-one states currently allow the use of the .220 Swift for deer hunting. The gun can fire a 50-grain (about a tenth of an ounce) .224-diameter bullet at a velocity approaching and even exceeding 4,000 ft/sec. Depending on the cartridge, muzzle velocities as high as 4,110 ft/sec are achievable with this weapon. It is known to be effective against deer at ranges of 200 yards. The barrel is long at 26 inches, but the gun has surprisingly little recoil, given its high power. A qualified marksman firing the .220 Swift can consistently shoot five shot groupings within a half-inch-diameter circle on a target positioned at a range of 100 yards. This weapon, firing a frangible round, seems to meet all the essential criteria, namely, an exceptionally accurate high-speed, small-caliber rifle.

But how does the Winchester .220 Swift stack up when confronted by the hard math of momentum conservation? Analysis of the Zapruder film shows that Kennedy's head recoiled a distance of 8.5 inches in 0.4 seconds, a velocity of 0.54 meters/sec.[3] The mass of Kennedy's head is not known, but Kennedy was a large man and his head likely weighed near the top of the human range for men, as much as 14 lbs., or 6.36 kilograms. Using this value, the momentum of his recoiling head is calculated by multiplying the mass of his head by the speed of its motion to the rear, or m × v (mass times velocity),

Winchester .220 Swift

Figure 21: The Winchester .220 Swift rifle. Introduced in 1935 as a superspeed centerfire, the Swift can send a 50-grain round downrange at speeds exceeding 4,000 ft./sec. Courtesy of Linda Alvey-Chambers.

expressed in mathematical symbols. This works out to be 3.43 kg-m/sec in MKS, (meter-kilogram-sec) units, a favorite standard measurement system of physicists.

The velocity of a bullet fired from a .220 Swift rifle, with losses due to air friction, is about 3,900 ft/sec, or 1,189 m/sec. The mass of its .224-caliber bullet is 50 grains, where one grain equals 65 milligrams, yielding a mass for the bullet of 3.24 grams. The bullet's momentum is therefore given by its mass times its velocity, or 3.85 kg-m/sec. If the bullet impacted Kennedy's head from an elevated angle, as would be the case if it were fired from the grassy knoll, for instance, then the component of momentum in the direction of recoil is reduced by the cosine of the angle of impact from the baseline of the pavement, since the motion of Kennedy's head is constrained by his neck and body. That is, we can only use the component of momentum of the bullet that is incident essentially along the horizontal. This is because of the constraints that Kennedy's anatomy would have placed on his range of motion. For a 15-degree angle of bullet incidence above the level of the street, it is necessary to multiply by cos 15°, or .966, to determine this component, yielding an incident momentum of 3.72 kg-me/sec for the incoming round that drives Kennedy's head backward.

This value is remarkably close to, and slightly above, the recoil momentum of Kennedy's head that was calculated above to be 3.43 kg-m/sec. The incoming bullet had some momentum to spare, allowing not only for the head recoil but for the momentum of the recoiling fragments as well. One large piece of Kennedy's scalp and skull ended up on the rear of his limousine. This is what Mrs. Kennedy was trying to retrieve when she climbed out of her seat onto the trunk.[4] This fragment, known as the Harper fragment, was ultimately found in the street, 25 feet to the rear of the location of Kennedy's limousine at the time of the fatal shot and recovered by medical student William Harper. As discussed previously, this skull piece was turned over to his uncle, Dr. Jack Harper, who analyzed it and concluded that it was parietal bone from the side of Kennedy's head.

This analysis is highly significant because it leads us to a conclusion completely consistent with the Zapruder film. The side of Kennedy's head was impacted and fragmented, not the rear (occipital) or the top of his skull. The skull fragment found in the street was from the side of his head (parietal), a determination based on striations and canal structure within the bone. "Parietal bone is characterized by a relatively smooth (excluding vascular grooves) inner surface, mild curvature, and relatively uniform thickness. In contrast, occipital bone is characterized by major variations on its internal surface (i.e., many different bumps and grooves from various things), much greater curvature, and substantial variation in thickness. Simply put, occipital bone doesn't look like the fragment from Kennedy's skull, but parietal bone does."[5]

The precise momentum of this fragment is unknown because it isn't visible on the Zapruder film. Assuming the fragment did not eject at a high angle from the horizontal, its velocity can be estimated by using the time it would have taken to fall a couple of feet, the distance from Kennedy's head to the limo body. Objects fall at a rate of 32 ft/sec^2, so the fragment would have fallen two feet in about a third of a second. Since the portion of his scalp landed about five feet in back of Kennedy, the rearward horizontal velocity is estimated to be about 15 ft/sec. Given its dimensions, a triangular shape about 7 cm by 5.5 cm on a side, an average thickness of 6 mm, the fragment's mass was on the order of about 20 grams. Given its weight, the momentum of the skull fragment was 0.1 kg-m/sec. Summing the momentum of this fragment with the recoil momentum of Kennedy's head, a value of 3.53 kg-m/sec is obtained. This number is in good accord with the momentum of the suspected bullet, calculated at 3.72 kg-m/sec for a .220 Swift round, leaving momentum to spare for other smaller fragments and some small angular momentum rotation of Kennedy's torso about his waist (the movement appears to be primarily torsional, or twisting about his waist).[6] This close agreement validates the deductions made at the beginning of the analysis, that the bullet was a .224-caliber, 50-gram frangible round traveling at 3,900 ft/sec. This identifies the Winchester .220 Swift as the assassin's murder weapon.

Examining one of the key starting assumptions, namely, that

gravity was not a major factor in the head recoil, requires use of the second conservation law, conservation of energy. This analysis is a little more complex and requires some more difficult math. Kennedy's head moved through a small arc as it pivoted on his neck and his shoulders, going from a listing angle of about 20 degrees to vertically upright. The center of mass of Kennedy's head rose at most a few inches through this arc. Using the formula mgh (mass times gravity times height) for the gravitational potential energy, where m is the mass of Kennedy's head, g is the acceleration of gravity (9.8 m.sec/sec), and h is the change in height (5 cm) of his head through the motion, a potential energy change of 3 joules (another MKS unit) is found to be due to gravity. However, the kinetic energy of the bullet, given by the formula $\frac{1}{2}mv^2$, was about 2,289 joules for the .220 Swift. Therefore, the energy of the bullet far exceeded the gravitational potential energy required to lift Kennedy's head, which is just another way of saying that because Kennedy's head was supported by his neck and body, the effect of the force of gravity was negligible. This calculation quantifies the starting assumptions and validates the analysis.

An analogy would be a boxer striking another boxer in the jaw or side of the head. The force from the punch would propel the head backward in recoil. The muscles of the neck are relatively weak and cannot withstand the force of a heavy blow from a punch. The same thing happens if a bullet strikes the skull from the side and does not penetrate through it. The bullet's momentum causes the head to recoil backward in the same way that it would in response to a left hook thrown by a boxer from the target's right front side.

Further use of the energy formula to determine the kinetic energy of Kennedy's recoiling head, $\frac{1}{2}mv^2$, provides a value of about a joule. But the energy of the incoming bullet was 2,289 joules. Where did all the energy go? Like Dennis's blocks, it simply went into other forms, some apparent, some hidden. Part of the energy of the bullet went into creating the high-speed jetting of blood and tissue in frame Z313. Some of it, 3 joules, went into lifting Kennedy's head against gravity as calculated above. The bulk of the energy, however, tragically, was deposited in Kennedy's brain. As the bullet and skull fragments penetrated his dura, the outer layer of the brain inside his skull, they transferred energy into Kennedy's neural tissues, breaking physical and chemical bonds and producing a rise in temperature. The kinetic energy of the fragments was converted into mechanical and thermal energy inside Kennedy's brain, obliterating a lifetime of thoughts, dreams, ambitions, hopes, and plans and leaving a lifeless lump of useless tissue. It was this violent deposition of energy directly into his head, over two-thousand joules (equivalent to a small explosive charge), that ended Kennedy's life.

This is an internally consistent self-contained model. All deductions and assumptions converged on agreement. Momentum conservation provides complete consistency with the initial deductions, and as consistency is the yardstick of truth, it is entirely reasonable to conclude that Kennedy was killed by a Winchester .220 Swift rifle firing from the front side, not the rear, of his head.

To summarize, the single-bullet theory failed when subjected to scientific scrutiny. Based on timing of the shots observed on the Zapruder film, failure of the single-bullet theory means that a second shooter was present in Dealey Plaza. Applying the law of conservation of momentum to the recoil of Kennedy's head at the instant of bullet impact observed on the Zapruder film (frame Z313), we conclude that the bullet must have impacted Kennedy from the right front side and remained inside his head. This forced Kennedy's head to absorb the recoil momentum from the bullet. Mathematical analysis of this recoil momentum shows it to be a very close match for the momentum of a .223-caliber round fired from a Winchester .220 Swift rifle. This result is consistent with the medical reports of copious bullet fragments discovered in Kennedy's brain and with the lack of evidence for an exit wound of any kind on Kennedy's head. Therefore, the shot that caused the head recoil exhibited in frame Z313 could not have been fired from behind Kennedy by a large-caliber, medium-velocity military rifle like the Mannlicher-Carcano, but instead originated from a high-velocity, small-caliber rifle firing from Kennedy's right front side.

Although alternative explanations have been put forward to

account for the head recoil, all suffer from serious deficiencies. The hypothesis that the head recoils due to the jetting of blood and tissue, like a macabre rocket, requires that the ejecta, because of its low mass, travel supersonically. If the jet of blood and brain matter was moving supersonically, one wouldn't see it on the slow-speed Zapruder film, any more than viewers can see the high-speed bullet on it. The idea that a listless Kennedy should spasm backward at the precise instant of bullet impact is exceedingly improbable. The motor area of the brain that controls movement of the hips and body trunk is located at the top of the brain, in the precentral gyrus, not on the side (parietal area) or rear (occipital) of the head. A bullet impact on the side or rear of the head could not trigger such a neurological spasm; and it certainly could not do so instantaneously with impact since nerve conduction needed to cause the spasm travels at finite speeds. These explanations are equivalent to adding epicycles (nested circles) to planetary orbits in order to salvage the Earth-centered model of the solar system. According to Occam's razor, the simplest explanation that fits the facts is almost always the right one.

Recall that momentum is a vector, having both a magnitude and a direction. The magnitude of the momentum vector for the bullet is in close agreement with the magnitude of the momentum vectors of Kennedy's recoiling head and fragments. But the Zapruder film records Kennedy's head recoiling back and toward Mrs. Kennedy, who is seated beside him to his left. The angle of the recoil then was therefore about 45 degrees with respect to the axis of the limousine body. Since momentum, both its magnitude and direction, must be conserved, the direction of the momentum of the incoming round must have been from this same angle relative to the limousine body. A bullet fired at an angle of 45 degrees to the limo axis traces a line directly back through the infamous grassy knoll. This is where the fatal head shot originated.

It doesn't matter if anyone saw or heard shouts coming from the grassy knoll. It doesn't matter if anyone saw a shooter in this location or not. Application of the incontrovertible laws of physics establishes that the bullet came from the direction of this site. In particular, tracing

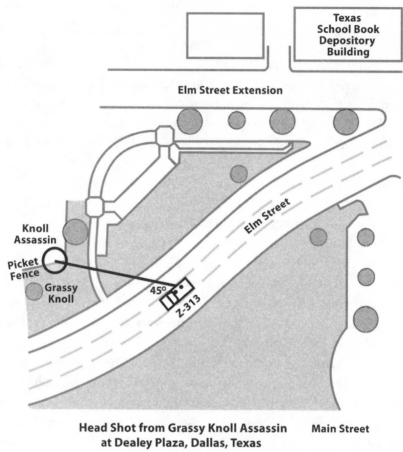

Head Shot from Grassy Knoll Assassin   **Main Street**
at Dealey Plaza, Dallas, Texas
November 22, 1963

Figure 22: Tracing a line back from Kennedy's limo at a 45 degree angle places the shooter behind the picket fence on the grassy knoll. Courtesy of Linda Alvey-Chambers.

a 45-degree angle back from the limousine at the time of the fatal shot places the assassin behind the west side picket fence on the grassy knoll.

A professional sniper using a high-tech weapon like the .220 Swift is apt to be adept at concealment. That's part of the art of the assassin. The grassy knoll area has many good concealment options, like the wooden picket fence shaded by a grove of trees. The fatal shot could have been fired from a location there or well behind it, so long as the shooter had a clear line of sight to his target. It may be that the shot was fired from a great distance at the exact moment that Kennedy's limo came into the sniper's line of sight. The most likely scenario was that the assassin was concealed in some manner behind the fences of the knoll area, firing an almost point-blank shot from a distance of 30 yards, a range from which he could not miss.

Some witnesses at Dealey Plaza reported seeing smoke emanating from the grassy knoll. Modern ball and tubular powders, bullet propellants, are designed not to produce smoke, that is, particulates, during combustion but instead generate almost exclusively gases upon deflagration, a reaction that is not quite as violent as an explosion. However, these powders are not entirely smokeless. Further, the round that struck Kennedy was frangible, and therefore likely a custom round prepared specially by the assassin. In this process, a hollow cylindrical shell casing would be loaded with a propellant, typically a ball powder. Then the lead bullet would be inserted into the shell casing, which would

Figure 23: Mary Moorman's photo taken moments after the fatal head shot. The box shows the projected location of the assassin based on momentum conservation.

then be "crimped" tightly around the projectile using a press-type machine. However, black powder, which produces copious smoke, could also have been used as the propellant in this process. An advantage of black powder is that, unlike modern, mostly smokeless ball powders, it isn't necessary to weigh it precisely before loading it into the custom-made bullet cartridge. This facilitates preparation of custom ammunition. Therefore, a rifle firing from this position could have potentially produced the copious smoke observed by witnesses on the grassy knoll.

Figure 23 shows the Polaroid photograph taken by Mary Moorman moments after the final, fatal shot struck Kennedy. Kennedy's head had clearly recoiled backward from the shot by the instant of this photo-

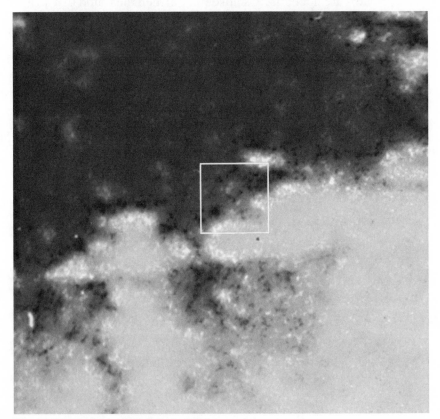

Figure 24: Enlargement of box in figure 23, the Mary Moorman photo.
Some claim that the image of a face is apparent (highlighted inside the box).

graph. The area above the wooden picket fence is highlighted by a box at the location where the assassin is determined to have fired this shot using the conservation law for momentum.

Figure 24 shows a computer-enhanced enlargement of the box. Some documentarians have claimed that the image of a face (highlighted within the box) is apparent at this location behind the west-side fence.

While the possibility exists that the apparent image is the face of a man standing behind the fence, and by implication, an assassin, the evidence for this is insufficient to meet the rigorous scientific standards of this analysis.

Although consistent conclusions can be reached concerning the final, fatal shot, the origin of the shot that penetrated Kennedy's throat is less clear. I am convinced, though, that the throat wound was in fact what the Dallas doctors said it was: an entrance wound. Some have argued that the bullet would have had to penetrate the windshield to hit Kennedy in the throat from the front. However, this need not have been the case if the shot was fired from an elevated position, for instance, from the Triple Overpass bridge overlooking Dealey Plaza, the ground next to it where a clear shot was open from the front side, or from the west-side fence position on the grassy knoll. The Dallas doctors were experts at the treatment and evaluation of gunshot wounds because they were trauma room physicians; many emergency trauma patients seek treatment because of gunshot wounds. Although no bullet was found by a cursory probe of the neck wound at Parkland Hospital, this round may also have been frangible or of small caliber, penetrating to the spine without exiting the back of the neck.

Some have conjectured that this round was an "ice" bullet laced with a paralytic agent that subsequently melted inside Kennedy's throat. An "ice" bullet would have required a dry ice dewar and could only have been loaded just prior to use, adding unwelcome complications for a concealed sniper trying to stay that way. As highlighted in chapter 2, the single-bullet "theory" is not viable and the expert conclusions of the Dallas emergency room physicians that the throat wound was an entrance wound must be assigned the weight of credibility.

The weapon that fired the final shot at Kennedy was a Winchester .220 Swift, not a Mannlicher-Carcano rifle. The final, fatal shot was fired not from behind Kennedy but was instead discharged from the right front side of his vehicle, from a location behind the west-side wooden picket fence on the grassy knoll, as demonstrated by application of the well-known and time-tested laws of physics to the images on the Zapruder film. Although the throat wound could have been inflicted while Kennedy's limo was at the location of the Stemmons Freeway sign by the .220 Swift that fired the fatal shot, it is also conceivable that it was caused by a third rifle firing from another location in front of Kennedy.

The ramifications of these conclusions are far-reaching, because there is overwhelming evidence for shots coming from a second location, the Texas School Book Depository. Many witnesses, including FBI agents, both heard and saw shots fired from the sniper's nest on the sixth floor of the depository. Governor Connally was hit by at least one shot, possibly two, from behind. Bullet fragments from these wounds were found in the front seat of the limo on the floor and were matched to the Mannlicher-Carcano rifle found on the sixth floor. Spent steel casings were also found in the sixth-floor sniper's nest. Everyone agrees that at least some shots must have been fired from this location.

All told, there is evidence for shots coming from behind Kennedy, fired from a military rifle, and convincing evidence for at least one shot originating from the front side, fired from a different weapon entirely, the .220 Swift, a high-tech assassin's rifle. Lee Harvey Oswald was seemingly capable of many things. He was accused of killing President Kennedy in Dealey Plaza and of murdering Officer Tippet on the other side of town within a forty-five-minute period. He had defected to Russia and came back again during the height of the cold war. But even Oswald could not have been in two places at once, firing two different rifles from two distinct locations within a six-second period. The probability of two autonomous assassins, acting wholly independently, firing at Kennedy at the same place at the same time is vanishingly small. Therefore, one must conclude from analysis of the dynamics exhibited on the Zapruder film that, with high confidence, Kennedy was killed in a crossfire involving multiple assassins.

This is exactly the same result obtained by the House Select Committee on Assassinations in 1978, based on the acoustic analysis of the Dictabelt recording taken during the assassination. Therefore, multiple independent instrumented records taken at the time of the event are in precise agreement that the last shot was fired by a rifle from the grassy knoll. With a second independent analysis confirming the result, the confidence in the conclusion compounds because not only must both records be wrong to disbelieve the outcome, but both records must be wrong in exactly the same way. The probability of this happening by chance is astronomically low. This is why scientists love multiple independent instrumented recordings. The probability of both recordings showing errors or instrumental artifacts in the same place at the same time in a manner consistent with a real physical phenomenon is extremely small. A second consistent independent instrumented record provides an enormous leap in confidence over just one such record.

As an example, if only a 96 percent probability of correctness is assigned to the Zapruder film analysis using momentum conservation (a low estimate in my opinion), the same value as the reliability of the acoustic analysis, one readily calculates that there is only a 0.16 percent chance (4 percent times 4 percent) that *both* analyses are erroneous. That translates to a 99.84 percent chance that at least one of the analyses is accurate and therefore a 99.84 percent chance that the final shot was fired from the grassy knoll. This value is certainly beyond reasonable doubt.

But the real odds of both results being wrong are much, much worse than this. Because not only must they both be wrong, they must both be wrong in precisely the same way. Now the probabilities become astronomical. Considering the length of the Dictabelt recording, roughly a thousand seconds, and the total possible places for a shooter in Dealey Plaza, at least a thousand, the probability of both records showing a rifle shot at the same time and the same place is one in a million. Multiplying this probability by the approximately one-in-a-thousand chance that both records are erroneous gives you an astronomical one-in-a-billion chance that both instrumented analyses could give identical false signals. This level of confidence is on a par with the accuracy of DNA testing. Clearly, a second rifle fired the fatal head shot in Dealey Plaza.

*　　*　　*

Still not convinced? Then I suggest you go down in your basement and test the laws of physics for yourself. Science is the ultimate "show me" adventure. If you don't believe something, do the experiment yourself. That's the beauty of science. You never have to accept anything on faith; you can always do your own experiment. If you don't believe that water boils at 100 degrees Celsius at sea level, put a thermometer in a pot of water and heat it. If you don't believe that different masses fall at the same velocity, drop a marble and a pool ball off your roof and see which one hits the ground first. If you don't think that oxygen is essential for combustion, place a candle under a glass and see how long it burns.

I encourage you to test momentum conservation for yourself. Use pool balls, bowling balls, baseballs, or my favorite, the air hockey table. An air table provides a virtually frictionless surface and is an ideal medium to test the laws of motion. Try impacting one disk on another across the table. After colliding, both disks move forward but at reduced velocity since the momentum must be apportioned between them. Keep in mind that momentum is a vector, so if you strike one disk into another through a glancing blow, the disks will go off in different directions but their momentum arrows will be added together, when lain head to tail, to equal the momentum vector of the incoming disc. If you like, tape the event with your camcorder and play it back at half speed to measure the velocities and directions accurately.

You can try a similar experiment with pool balls, but be careful to strike the cue ball dead center with the cue stick; otherwise you will introduce frictional forces with the felt of the table. This is the trick that some unscrupulous television producers have used in an attempt to convince people that Kennedy was shot from behind. They place a skull on a ladder and shoot the bottom of the skull, causing it to roll backward, due to friction with the ladder, and drop down off the back. Rolling friction is a different scenario than Kennedy's situation. His head was on a pivot, not rolling on a table. The complications of rolling are evident when you put spin on the cue ball by hitting it off center, causing it to career at an angle, even leap backward, after impacting another ball by virtue of the imparted spin interacting with the felt of the table through friction.

Having convinced ourselves of the accuracy of the conservation laws and having seen their application to the Kennedy assassination, only one conclusion is plausible: the final, fatal shot struck Kennedy from his right front side and was fired from a weapon exactly like the Winchester .220 Swift, not a Mannlicher-Carcano. Kennedy was therefore killed in a crossfire involving multiple assassins. Careful application of the time-tested laws of physics has led to this single result, the implications of which are profound.

Science is the ultimate test of truth. Lies and hypocrisy don't live long in the arena of modern scientific inquiry. It is fitting that the space program Kennedy championed should ultimately provide the methodology for resolving the riddle of his death: the physics of rocket science. With the identity of the murder weapon and the origin of the fatal shot established, the chances of unraveling the tangled threads of the Kennedy assassination, the Gordian knot of murder mysteries, increase exponentially. The law of nature is Alexander's sword.

CHAPTER 10

WHY IT MATTERS

Those who cannot remember the past are condemned to repeat it.
—George Santayana

He who controls the present controls the past. He who controls the
past controls the future.
—George Orwell, *1984*

The Civil War ended with the assassination of the president who had
won it. Abraham Lincoln, the first American president to be assassinated, was killed by a Confederate sympathizer as part of a larger plot,
which involved a simultaneous, although nonfatal, attack on Secretary
of State William Seward, and an unfulfilled attack on Vice President
Andrew Johnson. After John Wilkes Booth had placed his derringer
behind Lincoln's left ear and pulled the trigger, he exclaimed, "*Sic
semper tyrannis!*" (Ever thus to tyrants!). He then hurdled the rail of Lincoln's box to the stage, leaving the president bleeding from a mortal
head wound and in the arms of his shocked wife, Mary Todd. Booth,
summoning his renowned dramatic skills, cried out in the crowded theater, "The South is avenged!"[1]

The response to Lincoln's death was largely sympathetic. Indeed,
our knowledge of his grand funeral procession from Washington, DC,
back to Springfield, Illinois, where he had honed his political skills as a
lawyer and legislator, demonstrated the deep emotional pain that
Booth's act had caused. Even Robert E. Lee, the defeated general of the
South, lamented the tragedy of Lincoln's death. Other Southerners

were not so sympathetic. Some celebrated Booth as a bold avenger of Southern pride. As Southern diarist Kate Stone wrote, "All honor to J. Wilkes Booth. What torrents of blood Lincoln has caused to flow, and how Seward has aided him in his bloody work. I cannot be sorry for their fate. They deserve it. They have reaped their just reward."[2]

Still, for the restored Union, whether from the vantage point of a vengefully celebrant Southerner or an appalled, grieving Northerner, there were almost immediate answers to the questions surrounding the plot and the conspirators involved in murdering Lincoln. Newspapers and telegraph wires spread the news of the plot across the country and across the world. Citizens eagerly reviewed the material as reporters hashed out the facts based on eyewitness accounts and official statements given by investigators. They also undoubtedly insinuated their own opinions and beliefs about the complexity of the plot, those involved, and whether it was meritorious or evil.

Yet the public—those in the victorious North, those in the defeated South, and those who had been recently declared free men and women by decree of the federal government—achieved a sense of closure after the facts were corroborated. They knew who the responsible parties were, they knew the locations of where the plot was planned and executed, they knew the sensational details of how Booth and his conspirators accomplished their tasks, and they knew when the president was shot and how his body was carried across the street from Ford's Theater to a boarding house, where he died the next morning.

Booth even provided the stunned theatergoers of Ford's Theater, and in turn, the nation, with his reason for the plot: "The South is avenged!" His decreed motive is seared into the history of the United States, and his articulate command of the works of Shakespeare and his skilled stage presence give an overwhelming, if not also dubious, weight to the tragedy of his act.

We know the fate of Booth and a fellow conspirator as they made their escape south, toward land where Booth believed the people would be sympathetic to their acts. When he managed to get information on the response to the assassination, he was distraught to find he was mostly characterized as a monster, even in the South, where it was feared his act would reignite the feud between the former enemies and acts of retribution would swiftly befall the South for his slaying of Lincoln. Across the Union, Booth and his cohorts were wanted men. Warrant posters issued by the War Department offered a hundred thousand dollars for his capture.

Federal troops caught up with Booth, who was hiding in a tobacco barn in Virginia, southwest of the Potomac River. When he refused to leave the barn voluntarily, the troops set it afire. He was shot and dragged out of the barn, where he died while the sun began to rise over the Virginian plantation fields. In his journal Booth had written, in reference to Lincoln, "Our country owed all her troubles to him, and God simply made me the instrument of his punishment." Within three months, four of Booth's co-conspirators were hanged, including their landlady, Mary Surratt, who was executed over a public outcry questioning her role in Lincoln's murder. A fifth conspirator died in prison.[3]

Lincoln's death occurred a scant six days after Lee surrendered in Appomattox, Virginia, on April 9, 1865. Despite the fluctuations in his popularity as president, he had managed to win reelection in 1864 and retain his power as commander in chief during the remainder of the Civil War. He had delivered the Gettysburg Address, commemorating the untold numbers of war dead and declaring they had not died in vain. With his Emancipation Proclamation, he ushered in the Thirteenth Amendment, which officially abolished slavery and involuntary servitude. Few presidents have the list of accomplishments that Lincoln achieved, but few presidents have faced obstacles like Lincoln. When Booth's bullet silenced Lincoln forever, the nation lost a confirmed hero who had accomplished an unprecedented number of missions and who had used his rhetorical and political skill to hold a young, foundering nation together. The disclosure of the facts in his death provided not only closure to those millions who mourned the fallen president—and even those who once swore him to be an enemy—but the disclosure of these facts also mattered because our history matters.

Humans have the unique ability not only to remember events for sustained periods of time, but also to record those recollections. We record history not only for posterity, or because of our neurological craving for storytelling, or for politics or apologetics. We record history because, much like the DNA that provides instructions for us, our history provides instruction on how we founded our country, fought with other countries, succeeded, failed, and moved on. It yields an undeniable legacy instrumental in guiding our future. This history includes the most heroic as well as the most loathsome deeds committed under the auspices of the United States government and its citizens.

John F. Kennedy's death was a tragic alteration in the history of the United States, a paralyzing shock that stopped everyday life cold in late November 1963. Unlike Lincoln, who had managed to restore the Union among other enshrined moments in our nation's history and who had cemented his legacy a thousand times over, Kennedy was cut down in midstream, having just managed to reside in power for a thousand days. Lincoln's promise had been largely fulfilled. Kennedy's promise remains in question. But also unlike Lincoln, and every other assassination and assassination attempt that has taken place against a sitting American president, the facts of Kennedy's death are still in such dispute that the paralyzing shock of that November day is still there, and only failed attempts to soothe the pain have been provided.

Imagine how a true inquiry into the events surrounding Kennedy's death, the crime of the twentieth century, a shift in the course of our nation in the midst of confrontation between the superpowers, will likely provide us with vital historical knowledge. The untimely and perilous change in leadership effected by Kennedy's death heralded the turmoil and counterculture of the 1960s, a decade of war and internal strife, a clash of generations, and a precipitous loss of faith in government. We must uncover what really happened to our nation's leadership in that era.

Throughout human history, leaders have had tremendous influence on the cultures and nations they rule. This is consistent with our primate roots and is borne out through studies of lower primates. When baboons, a ground-dwelling type of monkey, were studied in the wild, it was observed that the behavior of the troupe had a great deal to do with the temperament of the alpha male. When an aggressive alpha male was put in charge of the troupe, the whole group went out and attacked neighboring troupes. When a more docile alpha male took over, the troupe stayed within its normal range, refusing to venture into neighboring areas and invite conflict.

This behavior is strikingly similar to that of human societies, regardless of the degree of civilization or technological proficiency. When a Genghis Kahn, Adolf Hitler, or Napoleon Bonaparte becomes the alpha male, the human troupe they lead ventures outside its range and attacks its neighbors. This primate behavior was a major source of pestilence for the Roman empire, which was plagued by generations of invading hordes led by the likes of Alaric, Odoacer, and Attila the Hun. When Kennedy was replaced by a more aggressive alpha male, Lyndon B. Johnson, America became more embroiled in a war halfway around the world, an event that would almost certainly not have happened had Kennedy continued his presidency.

History is in a sense defined by leaders. Our libraries are filled with testaments to the wills and acts of untold numbers of leaders, elected or crowned. Any reasonable review of history exhibits the pronounced effect that sovereigns have on the life and culture of their societies. Many historians cite Rome under the five good emperors—Nerva, Hadrian, Trajan, Antoninus Pius, and Marcus Aurelius—as some of the best times human civilization has ever known. Of course, it might not be hard to live as a Roman citizen in second-century Rome with its public facilities, staged events, and abundant slave labor. Will Durant writes, "The policies of Nerva and Trajan liberated the suppressed mind of Rome, and gave to the literature of their reigns a note of fierce resentment against a despotism that had gone but might come again."[4] Indeed, even if our forefathers decried the monarchic rule of Great Britain, they esteemed many of the principles and programs carried out by the Roman Republic and also later by the imperial rulers who judiciously employed their powers.

If Rome thrived under the astute leadership of Hadrian and

Marcus Aurelius, it suffered mightily under Caligula and Nero. Sadly, for every good Roman emperor, there were ten bad ones. Some were worse than others. Because there was no ballot vote every four years to determine whether the people were happy with their head of state, the Romans had to conceive of other ways to deal with bad emperors. This, of course, meant assassination. Over the course of their history, the Romans had perfected the art of assassination to both change national direction as well as to consolidate power.

Caligula was murdered by a tribune of his own guard, Cassius Chaerea, in a secret passage of his own palace. "On that day," the historian Dio tells us, "Caligula learned that he was not a god."[5] Caligula's own father, the successful and popular general Germanicus, was himself likely murdered by his uncle Tiberius, Caligula's predecessor, by poisoning, in order to become emperor himself. Nero committed suicide with the "help" of his freedman, Epaphroditus, after learning that the Senate had declared him a public enemy and condemned him to death. But despite his excesses, "[m]any more mourned for him, for he had been as generous to the poor as he had been recklessly cruel to the great."[6] The imperial age in Rome was ushered in by the assassination of Julius Caesar, ironically in a tragically misguided attempt to prevent a single individual from gaining too much power. After Caesar's death, the conspirators went out among the crowd proclaiming how they had saved Rome from tyranny. "Emerging, the conspirators found an excited populace in the square; they tried to win it with catch words of Liberty and the Republic, but the dazed crowd had no homage for phrases so long used to cover greed."[7] Within two years, Caesar's boasting assassins had been hunted down and killed.[8]

A president has significant power to influence the course of the nation. The president of the United States has the authority to start wars, to make appointments, to sign legislation, to grant clemency, and to interact with foreign governments in addition to wide-ranging emergency authorities. In the nuclear age, these powers are estimable. Certainly, presidents have a vast and enduring affect on the history and culture of their times. The movements and decisions of a president are scrutinized like never before, in the age of expanding niche news markets and 24/7 coverage. In live speeches before Congress and the nation, the president has enormous influence on public perception and the national direction. Some historians have referred to Franklin D. Roosevelt's time in office as the "the imperial presidency," a reference to his expanded role during that era. Although US presidents are not monarchs, just as other leaders in history they shape and define the national image, they have an inordinate impact on the direction of the country, and they have a direct influence on the path of its history. Kennedy's legacy was to put men on the moon, a seminal event of the twentieth century and one of the greatest achievements of the human race. The 1969 moon landing not only pushed the United States to the forefront of scientific and technological achievement, it also bolstered the nation's image and self-confidence in a way that no other nation has ever been able to enjoy. The space program championed by Kennedy inspired a generation of scientists and engineers and led to the burgeoning growth of the scientific culture of the twenty-first century. Kennedy's all-too-brief thousand days had an inordinate impact on the technological and cultural development of our society. That is one undeniable reason why his death was so tragic. Despite what his vision was able to instill and inculcate in those who followed and served it out, the full promise of what his presidency might further have accomplished still lingers large in the American consciousness.

Without a true and meaningful understanding of the actual perpetrators and their motives involved in the assassination of John F. Kennedy we have no real hope of preventing something like it from happening again. The assassination of a future US president in the nuclear age places the nation, and the world, in peril. An honest inquiry into the death of this consequential historical figure at the height of the cold war is therefore likely to reveal crucial insights that will be important for our nation as we attempt to move forward and navigate our future in the complex and competitive world of the twenty-first century. We can't know unless we actually perform the inquiry.

Many have speculated that organized crime, foreign governments, or rogue elements within our own government played a role in Kennedy's assassination. I do not know. I must leave it to others more knowledgeable than I am to determine who killed Kennedy and why. Some historians may settle for mysteries. They leave it to conspiracy theorists to make their cases and fulfill their agendas.

But scientists don't settle for mysteries.

The where and the when is known. On the morning of November 23, 1963, President and Mrs. Kennedy arrived on Air Force One in Dallas, climbed into the presidential limousine with Governor and Mrs. Connally, and made their way through the policed streets of the city. As previously stated, I could speculate on who is responsible for his death, but that is outside the purpose of this book. Again, I could even speculate on what the motives were, but that again is outside the purpose of this book. What I can say with confidence is the how: There were multiple shooters firing at the president with at least two different rifles in Dealey Plaza. Scientific analysis shows beyond peradventure that more than one weapon was involved in Kennedy's death.

A second shooter in Dealey Plaza indicates a conspiracy with enormous confidence. It would take suspension of credulity to imagine that two assassins shooting at Kennedy at the same time and same place were not associated. The mathematics of probability and the coincidence in timing of the shots bears this out. Kennedy's presidency lasted roughly a thousand days. No one shot at him until November 22, 1963. If he were around other people, appearing in crowds, or sitting near a window for about 10 percent of this time, or a little over 2 hours a day on average, his potential exposure would be about 8,640,000 seconds. The probability is therefore 1 in 1,440,000 that both shooters would attempt to fire at him by chance in the same 6-second period. It is more probable that civilization will come to an end next year due to an asteroid impact than it is likely that multiple assassins in Dealey Plaza were firing at Kennedy independently without prior coordination.

This analysis certainly supports those who suggest a conspiracy to assassinate Kennedy. Clearly, the assassin in the sixth-floor window did not act alone. A second gunman using a second type of weapon was involved. The Winchester .220 Swift, or a weapon like it, was the ideal choice for this gunman because it was effective at killing Kennedy without injuring Mrs. Kennedy. If Mrs. Kennedy had been hit by a penetrating bullet or fragment from a large-caliber round from the front, the public could never have been convinced that the fatal shot came from Oswald's rifle in the rear. This means the weapon must have been specified in advance. Whoever made this choice chose well and did so with an overall knowledge of a more-involved plan to implicate Oswald as the sole assassin.

The idea of a "lone nut" effecting Kennedy's death was an easy concept to advance in 1963 given the history of assassination attempts in the United States in the twentieth century. In October 1912, former president and then current Bull Moose presidential candidate Theodore Roosevelt emerged from dinner at Milwaukee's Hotel Gilpatrick on his way to give a campaign speech. As he entered an open car, John Nepomuk Shrank stepped forward and fired a .38 revolver into Roosevelt's chest. Roosevelt would go on to deliver his speech that night, exclaiming at the end, "It takes more than one bullet to kill a Bull Moose."[9] Shrank was later found to be insane and was committed to a series of mental institutions in Wisconsin until his death in 1943. His motive for shooting Roosevelt was that President William McKinley's ghost had come to him in a dream, accusing Roosevelt of his assassination in order to attain the office himself: "Let not a murderer take the presidential chair. Avenge my death."[10] McKinley had been shot at point-blank range by an anarchist, Leon F. Czolgosz, in 1901. "I killed President McKinley because I done my duty," he said after his arrest.[11] Although Czolgosz's lawyers were convinced he was insane, they never presented evidence to prove it at his trial. McKinley's killer died in the electric chair on October 29, 1901, declaring before his death, "I killed the president because he was the enemy of the people, the good working people. I am not sorry for my crime."[12] On his way to a mental institution by train, one of his guards asked John Nepomuk Shrank if he liked to hunt. Shrank replied, "Only Bull Moose."[13] After the death of McKinley, the secretary of the treasury charged the Secret Service, an organization created by Lincoln to pursue counterfeiters, with the task of officially protecting the president.[14]

John Wilkes Booth thought that he would receive national fame, or at least be idolized in the South, upon assassinating Lincoln, and so openly proclaimed what he had done. Both Shrank and Czolgosz, true "lone nuts," confessed to their crimes.

Oswald never did. In fact, he denied it. He called himself a patsy. And, to add a further twist, he was slain while "safely" in police custody by a lone gunman. Again, historians and others can speculate on theories as to why Oswald was gunned down, but chances are we may never know for certain.

Is it possible to paint a true and viable scenario of the events that transpired in Dealey Plaza on November 22, 1963? To some extent, the answer is yes. It is clear from the acoustic, medical, and film evidence we have reviewed that a volley of shots was fired at Kennedy's motorcade. Those shots involved at least one large-caliber military-style rifle firing from the depository, a super-high-velocity small-caliber weapon firing from the grassy knoll, and possibly a small-caliber rifle shooting at Kennedy from behind from another location. This would explain Kennedy's back wound. He was struck by a small-caliber round that did not penetrate his body. This round then later fell out, or was removed in the "pre-autopsy." It is also possible that a small-caliber (.22) round hit Kennedy in the rear of the head, driving fragments into his brain. At least one shot, however—probably two—was fired from a super-high-velocity rifle from the grassy knoll, the Winchester .220 Swift. As many as two assassins may have fired from the depository while an additional assassin possibly fired from another location in Dealey Plaza, such as the Dal-Tex Building.

If the presence of three or more assassins in Dealey Plaza strains credulity, consider that once it is clear that more than one assassin participated, it follows that the assassination of John F. Kennedy was the result of a carefully orchestrated plot. Once it is obvious that a conspiracy was under way, it is understandable that the use of several assassins enhanced the probability of success dramatically. As is evident from the medical evidence, not all shots struck the president; at least one shot struck Governor Connally, and another missed the motorcade completely and landed in the street. Success was therefore literally a hit-or-miss proposition. A reasonable analysis suggests that a single assassin would have had no more than a 50/50 chance of success. Based on this probability, however, three assassins would have provided an overall likelihood of success of 88.5 percent. Four shooters in Dealey Plaza would have provided a 94 percent chance of successfully assassinating the president. The consequences of failure would be frightening to any group conspiring to kill Kennedy. Had he lived, the full resources of the United States government, a true Hobbesian Leviathan, would have been unleashed to track down the conspirators. A shot and a miss would have been disastrous. It is not surprising then that the plotters would have used the principle of redundancy and employed a multiplicity of assassins to place the odds distinctly in their favor.

Indeed, Oswald had indicated during questioning that he knew of the presence of other weapons in the depository. In particular, he stated that he had seen a Mauser in the building.[15] Two days before the assassination, Warren Carter, an employee of the Southwestern Publishing Company, which occupied part of the second floor in the depository, had brought both a Mauser rifle as well as a .22-caliber rifle to the building to show his fellow employees. This was verified by numerous depository employees.[16] At a televised press conference the night of the assassination, Dallas district attorney Henry Wade stated that the rifle found in the depository was a 7.65 German Mauser.[17] This statement was actually corroborated by a CIA document on November 25, 1963, which stated "…employed in this criminal attack is a Model 91 rifle, 7.35 caliber, 1938 modification…the description of a Mannlicher-Carcano rifle in the Italian and foreign press is in error. It was a Mauser."[18] Thus, there was evidence for the presence of additional weapons in the Texas School Book Depository.

The assassin who fired the rifle shots from the grassy knoll was a professional, probably world-class. He likely hit Kennedy directly in the middle of his throat as his motorcade moved along Elm Street. He placed a second shot just off Kennedy's right ear, dead center on the side of his head, killing Kennedy cleanly with zero collateral damage. He hit no one else in Kennedy's limousine or motorcade. He escaped immediately, leaving no physical evidence of his presence. He left no gun, no spent cartridges, no fingerprints. He did not disturb the area in any traceable way. He has never been officially identified. This man was not someone like Oswald, but rather the type of person dramatized in Fred Zinnemann's 1973 film, *The Day of the Jackal*, based on the book of the same title by Frederick Forsyth. The extent to which the above analysis plays into current themes of conspiracy and the identification of Kennedy's assassins will ultimately be a matter for future historians to resolve.

There is a famous painting in Russia. In it, the muses of History present Stalin with the Books of History for his corrections. Is this what our government has done to the history of America? How will future historians view the "official" record of the assassination of John F. Kennedy? If a government can rewrite the history books with impunity, the nation takes one step closer to the world of George Orwell's novel *1984*. Indeed, in the mid-1970s, when the Zapruder film was finally aired publicly, investigation into the assassination was reopened. Although the House Select Committee on Assassinations determined there was a probable conspiracy, it never went beyond that to accuse any parties, besides the long-dead Oswald. It did, however, advocate that the Justice Department should reopen the investigation into Kennedy's death. This, of course, never happened. The government's official version was written by the previous commission, some of whose members wholly disagreed with their own sworn findings, and then reexamined and called into question a decade later, though no significant gains had been made in determining the party or parties responsible. It may not be the dystopian world of Orwell's characters who are fed lies daily, but it certainly is cold comfort for those who mourned the loss of the president and his promise that a mystery still surrounds his death. To deny Americans the truth of the assassination is to deny us our history.

WHY IT MATTERS

Many cite the Kennedy assassination as the day that America lost faith in its government. Indeed, the assassination and its aftermath have led to a rift in the American psyche, a gaping wound that eats at the national consciousness and must be healed. There is no greater balm than the truth. The minimum standard for a government to enjoy the good will of its people is honesty. No government can consider that it serves its people if it blatantly lies to those people. If America is continually denied the truth about the Kennedy assassination, a central event in the nation's history, the repercussions of which still affect us today, then the future of the nation is irreparably compromised. What you don't know will end up hurting you someday. In order for this nation to learn and grow from its history it must first know just what that history was. Only honesty will set the record straight, allow the wounds to heal, and restore confidence in our government.

Almost fifty years have passed since Kennedy's tragic death. The prospects of World War III stemming from the truth of the assassination are now remote. Serious damage to the nation's reputation as a result of a serious inquiry into Kennedy's death is also unlikely. In fact, I would argue that continued secrecy as to the true events surrounding Kennedy's death does significant harm to our national image; other countries likely look at the United States as a nation in denial of its formative past.

With the scientific evidence directly supporting multiple shooters in Dealey Plaza, it is time to reopen the investigation into Kennedy's death anew. Given sufficient motivation by the electorate, the present generation of leaders may be able to succeed in truly reclaiming our history where the past generation has, tragically, failed.

EPILOGUE

Following publication of his groundbreaking yet caustic *Dialogue concerning the Two Chief World Systems*, Galileo Galilei was ordered to stand trial on suspicion of heresy in 1633. Galileo's unrepentant advocacy of the Copernican system had finally gotten him into trouble. Although he had been careful to veil his true beliefs in his published works by placing his views in the words of characters in dialogues, he did not fool the Inquisition. But Galileo actually had proof of the falsity of an Earth-centered planetary system. He had seen the moons of Jupiter in his telescope—incontrovertible evidence that all heavenly bodies did not circle Earth. The Inquisition was not impressed: Galileo was found "vehemently suspect of heresy," namely, of having held the opinions that the sun lies motionless at the center of the universe, that Earth is not at its center and that it moves, and that one may hold and defend an opinion as probable after it has been declared contrary to Holy Scripture. He was required to "abjure, curse and detest" those opinions. He was ordered imprisoned, but the sentence was later commuted to house arrest. His offending *Dialogue* was banned; publication of his works was forbidden, including any he might write in the future. The founder of modern physics remained under house arrest for the rest of his life.

Standing up to authority takes courage. Many men and women have come forward over the years to challenge the edicts and pronouncements of authorities who pronounced, and still pronounce to this day, that Oswald acted alone in Dealey Plaza. This book gives a fresh voice to the men and women who reported hearing shots fired from the grassy knoll, because the physical evidence from analysis of the film and acoustic records now firmly supports them.

That empirical evidence consists of medical records, acoustic records, and film records. The Zapruder film lays the foundation for the unmistakable impression that the final shot struck Kennedy from the front. Careful scientific analysis of the medical, acoustic, and ballistic evidence bears this out and leads ineluctably to a single conclusion beyond argument: someone other than Oswald fired shots that killed Kennedy. A mystery therefore still surrounds Kennedy's death. What role did Oswald play, who were the co-conspirators, and who were the assassins?

The purpose of this book is not to prove who killed Kennedy, nor why. This work does not identify Kennedy's killers, nor does it pretend to elucidate the intricate internal politics of 1960s America. However, detailed scientific analysis establishes the involvement of multiple people in Kennedy's death. It is beyond coincidence that shots from two separate rifles, fired so closely in time, could be the work of two completely independent individuals who were acting exclusively on their own. Although I offer no opinion as to the identities, Kennedy's assassin must have been acting in concert with associates.

Kennedy talked often to his associates about the possibility that he might one day be struck by an assassin's bullet. He even confided in his wife that it was always possible that some nut with a gun could take a shot at him. But he didn't allow himself to worry about it. What Kennedy didn't anticipate as he rode into Dealey Plaza on November 22, 1963, was that a team of professional assassins lay in wait for him. This act changed the course of history.

NOTES

INTRODUCTION

1. Jim Marrs, *Crossfire: The Plot that Killed Kennedy* (New York: Carol & Graf, 1989), p. 356.

2. Vincent Bugliosi, *Reclaiming History: The Assassination of President John F. Kennedy* (New York: Norton, 2007), p. 177.

CHAPTER 1

1. Edward Jay Epstein, *Inquest: The Warren Commission and the Establishment of Truth* (New York: Bantam, 1966), p. xi.

2. Earl Warren, *The Memoirs of Earl Warren* (Garden City, NY: Doubleday, 1977).

3. FBI, Report on the Assassination of President John F. Kennedy, December 9, 1963.

4. Warren Commission, *Report of the President's Commission on the Assassination of President Kennedy* (Washington, DC: GPO, 1964), foreword, p. x.

5. Ibid., p. 476.

6. Epstein, *Inquest*, p. 9.

7. Ibid., p. 11.

8. Ibid., p. 12.

9. Epstein, *Inquest*, p. 60.

10. Vincent Bugliosi, *Reclaiming History: The Assassination of President John F. Kennedy* (New York: Norton, 2007), p. 346.

11. Warren Commission, *Investigation of the Assassination of President John F. Kennedy: Hearings before the President's Commission on the Assassination of President Kennedy* (Washington, DC: GPO, 1964), vol. 6, p. 244.

12. Ibid., vol. 17, p. 882.

13. Ibid., vol. 4, pp. 114, 128.

14. Ibid., vol. 2, p. 375.

15. Ibid., vol. 5, p. 86.

16. Michael L. Kurtz, *The JFK Assassination Debates* (Lawrence: University Press of Kansas, 2006), p. 30.

17. David Lifton, *Best Evidence* (New York: Macmillan, 1980), p. 23.

18. Epstein, *Inquest*, p. 105.

19. Ibid., p. 106.

20. Ibid.

21. Ibid.

22. Ibid., p. 107.

23. David R. Wrone, *The Zapruder Film: Reframing JFK's Assassination* (Lawrence: University Press of Kansas, 2003), p. 243. Citing tape transcripts released from the Lyndon B. Johnson Library in Austin, TX.

CHAPTER 2

1. Warren Commission, *Investigation of the Assassination of President John F. Kennedy: Hearings before the President's Commission on the Assassination of President Kennedy* (Washington, DC: GPO, 1964), vol. 2, p. 34.

2. Edward Jay Epstein, *Inquest: The Warren Commission and the Establishment of Truth* (New York: Bantam Books, 1966), p. 30.

3. Ibid.

4. Ibid.

5. Warren Commission, *Hearings*, vol. 14, p. 522.

6. Ibid., vol. 2, p. 169.

7. FBI supplemental report, January 13, 1964, p. 2.

8. Warren Commission, *Hearings*, vol. 2, p. 93.

9. Epstein, *Inquest*, p. 49.

10. Ibid., vol. 4, p. 135.

11. Ibid., p. 147.

12. Warren Commission, *Report of the President's Commission on the Assassination of President Kennedy* (Washington, DC: GPO, 1964), p. 116.

13. Warren Commission, *Hearings*, vol. 6, pp. 130–31.

NOTES

14. Ibid., vol. 2, p. 382.

15. Epstein, *Inquest*, p. 114.

16. *Hearings*, vol. 3, p. 406.

17. Vincent Bugliosi, *Reclaiming History: The Assassination of President John F. Kennedy* (New York: Norton, 2007), pp. 494–95.

18. Warren Commission, *Report*, p. 194.

19. Phillip J. Corso, *The Day after Roswell* (New York: Simon & Schuster, 1997).

20. "Top Ten Literary Hoaxes," *Guardian*, November 15, 2001, http://www.guardian.co.uk/books/2001/nov/15/news (accessed June 11, 2010).

21. Dick Russell, *On the Trail of the JFK Assassins* (New York: Skyhorse Publishing, 2008), p. 126.

22. Epstein, *Inquest*, p. 149.

23. Ibid.

24. Ibid., p. 61.

25. Ibid., p. 107.

26. Ibid., p. 81.

27. Ibid., p. 83.

28. Michael L. Kurtz, *The JFK Assassination Debates* (Lawrence: University Press of Kansas, 2006), p. 86.

29. Ibid., p. 50.

CHAPTER 3

1. Colin Tudge, *The Link: Uncovering Our Earliest Ancestor* (New York: Little, Brown, 2009).

2. Robert J. Groden, *The Killing of a President* (New York: Penguin, 1993), pp. 54–57.

3. Jim Marrs, *Crossfire: The Plot That Killed Kennedy* (New York: Caroll & Graf, 1989), p. 86.

4. Ibid., p. 70.

5. Ibid., p. 71.

6. Ibid., p. 72.

7. Ibid., p. 87.

8. Warren Commission, *Investigation of the Assassination of President John F.*

NOTES

Kennedy: Hearings before the President's Commission on the Assassination of President Kennedy (Washington, DC: GPO, 1964), vol. 6, p. 243.

9. *Dallas Morning News*, November 23, 1963.

10. National Archives, Basic Source Materials in Possession of Commission: Commission no. 87, folder no. 1, Secret Service control no. 66.

11. Ibid., pp. 81–85.

12. Lamar Waldron, *Ultimate Sacrifice* (New York: Carol & Graf, 2005), p. 719.

13. Josiah Thompson, *Six Seconds in Dallas* (New York: B. Geis Associates, 1967), p. 155.

14. Waldron, *Ultimate Sacrifice*, p. 721.

15. Marrs, *Crossfire*, p. 78.

16. Lamar Waldron, *Ultimate Sacrifice*, p. 722.

17. Marrs, *Crossfire*, p. 81.

18. Waldron, *Ultimate Sacrifice*, p. 724.

19. Marrs, *Crossfire*, p. 39.

20. Harold Feldmen, "Fifty-one Witnesses: The Grassy Knoll," *Minority of One*, March 1965.

21. *Texas Observer*, December 13, 1963.

22. Ibid.

23. Anthony Summers, *Conspiracy* (New York: McGraw-Hill, 1980), p. 82.

24. Mark Lane, *Rush to Judgment* (New York: Holt, Rinehart, & Winston, 1966), p. 30.

25. Warren Commission, *Hearings*, vol. 24, p. 217.

26. Lane, *Rush to Judgment*, p. 40.

27. Robert McNeill, ed., *The Way We Were: 1963, The Year Kennedy Was Shot* (New York: Carol & Graf, 1988), p. 195.

28. Warren Commission, *Hearings*, vol. 22, p. 648.

29. Ibid., p. 634.

30. Ibid., p. 632.

31. Ibid., p. 386.

32. Ibid., p. 648.

33. Ibid., vol. 7, p. 346.

34. Ibid., vol. 18, p. 759.

35. Thomas P. O'Neill, *Man of the House: The Life and Political Memoirs of Speaker Tip O'Neill* (New York: Random House, 1987), p. 178.

NOTES

CHAPTER 4

1. Joseph S. Weiner, *The Piltdown Forgery* (London: Oxford University Press, 1955).

2. Roger Lewin, *Bones of Contention: Controversies in the Search for Human Origins* (Chicago: University of Chicago Press, 1997).

3. Arthur Smith Woodward, *The Earliest Englishman* (London: C. C. Watts, 1948).

4. Kenneth F. Oakley and J. S. Weiner, "Piltdown Man," *American Scientist* (October 1955).

5. Weiner, *The Piltdown Forgery*.

6. Stephen Jay Gould, "Piltdown Revisited," in *The Panda's Thumb* (New York: Norton, 1980).

7. John Hathaway Winslow and Alfred Meyer, "The Perpetrator at Piltdown" *Science 83* (September 1983).

8. R. J. Beuhler, G. Friedlander, L. Friedman, "Cluster-Impact Fusion," *Physical Review Letters* 63, no. 1292 (1989).

9. Private communication with J. C. Walker, professor, Department of Physics, Johns Hopkins University.

10. Denis L. Rousseau, "Case Studies in Pathological Science," *American Scientist* 80, no. 1 (January/February 1992): 54–63.

11. Carl Sagan, *Cosmos* (New York: Random House, 1980), p. 185.

12. Ibid., p. 58.

CHAPTER 5

1. Gerald Posner, *Case Closed* (New York: Random House, 1993), p. 288.

2. J. A. Nicholas, C. L. Burstein, C. J. Umberger, and P. D. Wilson, "Management of Adrenocortical Insufficiency during Surgery," *Archives of Surgery* (1955/1971): 737–42. JFK is case 3.

3. David Lifton, *Best Evidence* (New York: Macmillan, 1980), p. 279.

4. Ibid.

5. Michael L. Kurtz, *The JFK Assassination Debates* (Lawrence: University Press of Kansas, 2006), p. 7.

6. United Press International "A" wire, 3:10 CST, November 22, 1963.

NOTES

7. M. T. Jenkins, "Statement concerning Resuscitative Efforts for President John F. Kennedy," November 22, 1963, http://mcadams.posc.mu.edu/jenkins2.gif (accessed June 11, 2010).

8. Joseph Riley, "Anatomy of the 'Harper Fragment,'" http://roswell.fortunecity.com/angelic/96/harper~1.htm (accessed June 11, 2010).

9. Jim Marrs, *Crossfire: The Plot That Killed Kennedy* (New York: Caroll & Graf, 1989), p. 363.

10. Lifton, *Best Evidence*, pp. 403–407.

11. Art Peterson, "Another Link in JFK Death?" *News Sun*, May 1, 1975.

12. Ibid.

13. Marrs, *Crossfire*, p. 368.

14. Lamar Waldron with Thom Hartmann, *Legacy of Secrecy: The Long Shadow of the JFK Assassination* (Berkeley: Counterpoint, 2009), p. 182.

15. Ibid., p. 183.

16. Ibid.

17. Kurtz, *The JFK Assassination Debates*, pp. 39–40.

18. Francis X. O'Neill and James W. Sibert, "Autopsy of Body of President John Fitzgerald Kennedy," FBI report dictated by FBI agents Francis X. O'Neill and James W. Sibert, November 26, 1963, p. 2.

19. Warren Commission, *Report of the President's Commission on the Assassination of President Kennedy* (Washington, DC: GPO, 1964), appendix IX, "Autopsy Report and Supplemental Report," p. 544.

20. Warren Commission, *Hearings before the President's Commission on the Assassination of President Kennedy* (Washington, DC: GPO, 1964), vol. 2, p. 353.

21. Marrs, *Crossfire*, p. 371.

22. O'Neill and Sibert, "Autopsy of Body of President John Fitzgerald Kennedy," p. 4.

23. Marrs, *Crossfire*, p. 371, quoting Sibert and O'Neill's report.

24. Lifton, *Best Evidence*, pp. 403–407.

25. William Matson Law, *In the Eye of History* (Southlake, TX: JFK Lancer Productions, 2004), pp. 132, 137.

26. Tom Wicker, "Kennedy Is Killed by Sniper as He Rides in Car in Dallas," *New York Times*, November 23, 1963.

27. Lifton, *Best Evidence*, p. 272.

28. Ibid.

29. AARC, Bethesda Autopsy Report, p. 3, http://www.aarclibrary.org/

NOTES

publib/jfk/wc/wcvols/wh16/pdf/WH16_CE_387.pdf (accessed June 11, 2010).

30. Lifton, *Best Evidence*, p. 274.

31. Ibid., p. 190.

32. O'Neill and Sibert, "Autopsy of Body of President John Fitzgerald Kennedy," p. 5, quoted in Marrs, *Crossfire*, p. 372.

33. Ibid., p. 2.

34. Marrs, *Crossfire*, p. 317.

35. Robert Groden, *The Killing of a President* (New York: Viking Penguin, 1993), p. 85.

36. G. Robert Blakey, private communication, May 19, 2010.

37. Edward Jay Epstein, *Inquest: The Warren Commission and the Establishment of Truth* (New York: Bantam, 1966), p. 46.

38. Ibid., p. 49.

39. Warren Commission, *Report*, Appendix IX, "Autopsy Report and Supplemental Report," p. 543.

40. Kurtz, *The JFK Assassination Debates*, p. 36.

41. Bonar Menninger, *Mortal Error: The Shot That Killed JFK* (New York: St. Martin's Press, 1992).

42. Discovery Channel, *JFK: Inside the Target Car*, November 16, 2008.

43. Ibid.

44. Lifton, *Best Evidence*, p. 580.

45. Francis X. O'Neill and James W. Sibert, "Autopsy of Body of President John Fitzgerald Kennedy," FBI report dictated by FBI Agents Francis X. O'Neill and James W. Sibert on November 26, 1963, p. 2.

CHAPTER 6

1. House Select Committee on Assassinations, *Report of the Select Committee on Assassinations of the US House of Representatives*, 95th Congress, 2nd session (Washington, DC: GPO, 1979).

2. Jim Marrs, *Crossfire: The Plot That Killed Kennedy* (New York: Carol & Graf, 1989), p. 529

3. Ibid.

4. Mary Ferrell Foundation, "Acoustic Evidence," www.maryferrell.org/wiki/index.php/Acoustics_Evidence (accessed June 11, 2010).

NOTES

5. House Select Committee on Assassinations, *Findings of the Select Committee on Assassinations in the Assassination of John F. Kennedy* (Washington, DC: GPO, 1979), section I.B., "Scientific Acoustical Evidence," p. 68.

6. Ibid., p. 69.

7. Ibid.

8. Ibid.

9. Ibid., p. 83.

10. Ibid., p. 70.

11. Ibid., p. 71.

12. Ibid., p. 72.

13. D. B. Thomas, "Overview and History of the Acoustical Evidence in the Kennedy Assassination Case," http://www.maryferrell.org/wiki/index .php/Essay_-_Acoustics_Overview_and_History (accessed June 11, 2010).

14. House Select Committee on Assassinations, *Findings*, Section I.B., "Scientific Acoustical Evidence," pp. 73–74.

15. Ibid., p. 74.

16. Ibid., pp. 75–78.

17. Ibid., p. 93.

18. Marrs, *Crossfire*, p. 536.

19. National Research Council, *Report of the Committee on Ballistic Acoustics*, prepared for Department of Justice, Washington, DC. Report no. PB83-218461, 1982.

20. Ibid., p. 34.

21. Ibid., pp. 18–31.

22. National Research Counsel, *Report of the Committee on Ballistic Acoustics*, prepared for the Department of Justice, Washington, DC, Report No. PB83-218461 (1982), p. 40.

23. Ibid.

24. Ibid., p. 52.

25. FBI record 124-10006-10153, letter from James Barger to G. Robert Blakey, February 18, 1983, pp. 1–3.

26. D. B. Thomas, "Echo Correlation Analysis and the Acoustic Evidence in the Kennedy Assassination Revisited," *Science & Justice* 41 (2001): 21–32.

27. Ibid., p. 31.

28. David R. Wrone, *The Zapruder Film: Reframing JFK's Assassination* (Lawrence: University Press of Kansas, 2003), p. 118.

29. Vincent Bugliosi, *Reclaiming History: The Assassination of President John F. Kennedy* (New York: Norton, 2007), p. 475.

30. Gerald L. Posner, *Case Closed* (New York: Random House, 1993), p. 329.

31. Wrone, *The Zapruder Film*, p. 133.

32. G. Robert Blakey, private communication, May 19, 2010.

33. Cyril H. Wecht, "Supplement to Addendum to Forensic Pathology Panel Report Submitted by Cyril H. Wecht, M.D. Memo to Robert Blakey, 10 August 1978," HSCA box 196, #010683.

34. D. B. Thomas, "Echo Correlation Analysis," p. 26.

35. R. Linsker, R. L. Garwin, H. Chernoff, P. Horowitz, N. F. Ramsey, "Synchronization of the Acoustic Evidence in the Assassination of President Kennedy," *Science & Justice* 45, no. 4 (October 2005): 207–26.

36. G. P. Chambers, J. E. Eridon, and K. S. Grabowski, "Upper Limit on Cold Fusion in Thin Palladium Films," *Physical Review B* 41, no. 6 (March 1990): 5388–91.

37. J. Fine Marton and G. P. Chambers, "Temperature Dependent Radiation-Enhanced Diffusion in Ion-Bombarded Solids," *Physical Review Letters* 61, no. 23 (December 1988): 2697–2700.

38. Jonathan Weiner, *Planet Earth* (New York: Bantam, 1986), pp. 15–17.

39. Ibid.

40. Abraham Pais, *Subtle Is the Lord: The Science and the Life of Albert Einstein* (Oxford: Oxford University Press, 1982), p. 511.

41. Albert Messiah, *Quantum Mechanics* (New York: Wiley, 1976), vol. 2, p. 544.

42. S. A. Goudsmit, "The Discovery of the Electron Spin," www.lorentz.leidenuniv.nl/history/spin/goudsmit.html (accessed June 11, 2010).

43. Ibid.

CHAPTER 7

1. Vincent Bugliosi, *Reclaiming History: The Assassination of President John F. Kennedy* (New York: Norton, 2007), p. 488.

2. Gerald Posner, *Case Closed* (New York: Random House, 1993).

3. Ibid., p. 256.

4. Jim Marrs, *Crossfire: The Plot That Killed Kennedy* (New York: Carol & Graf, 1989), pp. 58–59.

NOTES

5. Posner, *Case Closed*, p. 252.

6. Ibid.

7. Ibid., p. 257.

8. Ibid., appendix A, p. 481.

9. Ibid., p. 472.

10. Ibid., pp. 317–18.

11. Ibid., p. 410.

12. Ibid., p. 472.

13. Bugliosi, *Reclaiming History*, p. xxiv.

14. Ibid., p. xxv.

15. David R. Wrone, *The Zapruder Film: Reframing JFK's Assassination* (Lawrence: University Press of Kansas, 2003), p. 245.

16. Jim Moore, *Conspiracy of One: The Definitive Book on the Kennedy Assassination* (Ottawa, ON: Summit Publishing Group, 1990).

17. Bugliosi, *Reclaiming History*, p. xxxvii.

18. Ibid.

19. Ibid.

20. Ibid., pp. 489–90.

21. Ibid., p. 499.

22. Ibid.

23. Ibid., p. 500.

24. Wrone, *The Zapruder Film*, p. 212.

25. Bugliosi, *Reclaiming History*, p. 458.

26. Ibid., p. 475.

27. Posner, *Case Closed*, p. 329.

28. Ibid., pp. 329–30.

29. Warren Commission, *Report of the President's Commission on the Assassination of President John F. Kennedy* (Washington, DC: GPO, 1964), p. 557.

30. Wrone, *The Zapruder Film*, p. 215.

31. Warren Commission, *Hearings before the President's Commission on the Assassination of President Kennedy* (Washington, DC: GPO, 1964), vol. 4, p. 113.

32. Wrone, *The Zapruder Film*, p. 214.

33. Ibid., p. 229.

34. Ibid., p. 230.

35. Michael L. Kurtz, *The JFK Assassination Debates* (Lawrence: University Press of Kansas, 2006), p. 114.

NOTES

36. Bugliosi, *Reclaiming History*, p. 397.

37. Ibid., p. 1452.

38. Ibid.

39. Bonar Menninger, *Mortal Error: The Shot That Killed JFK* (New York: St. Martin's Press, 1992).

40. Discovery Channel, *JFK: Inside the Target Car*, November 16, 2008.

41. Bugliosi, *Reclaiming History*, p. 484.

42. Ibid., p. 486.

43. Ibid., p. 484.

44. Wrone, *The Zapruder Film*, p. 103.

45. Ibid.

46. Warren Commission, *Hearings*, vol. 6, pp. 294–95.

47. Eric Bland "Tech Snuffs JFK Controversy," *Discovery News*, http://dsc.discovery.com/technology/im/jfk-gary-mack-tech-snuffs.html (accessed June 16, 2010).

48. Feynman's Messenger Lectures, delivered at Cornell University in 1965, were collected by Bill Gates of Microsoft Research and can be played from his Web site Project Tuva at http://research.microsoft.com/apps/tools/tuva.

49. Bugliosi, *Reclaiming History*, p. 849.

50. Ibid.

51. Josiah Thompson, "The Cross Fire That Killed President Kennedy," *Saturday Evening Post*, December 2, 1967.

52. Menninger, *Mortal Error*.

53. *Weisberg v. ERDA and the Department of Justice*, Civil Action 75-226; Wrone, *The Zapruder Film*, p. 172.

54. FBI Lab report, March 6, 1964, 105-82555-2384.

55. Paraffin test records, 75-226 file, Weisberg Archives.

56. Marrs, *Crossfire*, p. 51.

57. Wrone, *The Zapruder Film*, p. 170.

58. Bugliosi, *Reclaiming History*, p. 1457.

59. Ibid., p. 107.

60. Ibid., p. 949.

61. Ibid., p. 864.

62. Ibid.

63. Ibid., p. 1489.

NOTES

CHAPTER 8

1. G. P. Chambers, H. Sandusky, F. Zerilhi, K. Rye, R. Tussing, and Jerry Forbes, "Pressure Measurements on a Deforming Surface in Response to an Underwater Explosion in a Water-Filled Aluminum Tube," *Shock and Vibration* 8 (2001): 1–7.

2. Mark Cotta Vaz and Craig Barron, *The Invisible Art: The Legends of Movie Matte Painting* (San Francisco: Chronicle Books, 2002).

3. David R. Wrone, *The Zapruder Film: Reframing JFK's Assassination* (Lawrence: University Press of Kansas, 2003).

4. Warren Commission, *Hearings before the President's Commission on the Assassination of President Kennedy* (Washington, DC: GPO, 1964), vol. 7, p. 570.

5. Wrone, *The Zapruder Film*, p. 19.

6. Josiah Thompson, *Six Seconds in Dallas: A Micro-Study of the Kennedy Assassination* (New York: B. Geis Associates, 1967), p. 102.

7. Warren Commission, *Hearings*, vol. 5, p. 352.

8. Wrone, *The Zapruder Film*, p. 20.

9. Ibid., p. 21.

10. Ibid., p. 23.

11. Affidavit of P. M. Chamberlain Jr., November 22, 1963, in Roland Zavada, "Analysis of Selected Motion Picture Photographic Evidence," Kodak technical report, September 7, 1998, study I, appendix.

12. Affidavit of Frank R. Sloan, November 22, 1963, in Zavada, "Analysis of Selected Motion Picture Photographic Evidence," study I, appendix.

13. Newseum with Cathy Trost and Susan Bennett, *President Kennedy Has Been Shot* (Naperville, IL: Sourcebooks, 2003), p. 164.

14. Ibid., p. 186.

15. Wrone, *The Zapruder Film*, pp. 121–41.

16. Philip H. Melanson, "Hidden Exposure: Cover-up and Intrigue in the CIA's Secret Possession of the Zapruder Film," *Third Decade*, November 1984, pp. 13–21.

17. James H. Fetzer, ed., *Assassination Science: Experts Speak Out on the Death of JFK* (Chicago: Catfeet Press, 1997).

18. Harold Weisberg, *Whitewash* (Hyattstown, MD: self-published, 1965); *Never Again!* (New York: Carol & Graf, 1995), pp. 301–305.

19. Wrone, *The Zapruder Film*, p. 123.

NOTES

20. Ibid., p. 124.

21. Zavada, "Analysis of Selected Motion Picture Photographic Evidence."

22. Joe Durnavich and David Wimp, "David Healy's Technical Aspects of Film Editing," home.earthlink.net/~joejd/jfk/zaphoax/healy.html (accessed June 16, 2010).

23. Roland Zavada, "Zapruder Film Hoax Response," September 23, 2003, http://www.clintbradford.com/fetzerfails3.htm (accessed June 16, 2010).

24. Ibid.

25. Ibid.

26. Michael L. Kurtz, *The JFK Assassination Debates* (Lawrence: University Press of Kansas, 2006), p. 106.

CHAPTER 9

1. Vincent Bugliosi, *Reclaiming History: The Assassination of President John F. Kennedy* (New York: Norton, 2007), p. 488.

2. Anonymous source.

3. Ibid., p. 489, footnote.

4. *Hearings before the President's Commission on the Assassination of President Kennedy* (Washington, DC: GPO, 1964), vol. 2, p. 140.

5. Joseph Riley, "Anatomy of the 'Harper Fragment,'" www.roswell .fortunecity.com/angelic/96/harper~1.htm (accessed June 16, 2010).

6. Depending on the degree of de-convolution of the backward body movement between twisting and rotation about the waist, which is difficult to determine from the film frames, the angular momentum of rotation about Kennedy's waist could be significant and would require additional momentum in the calculation. However, the choice of the mass of Kennedy's head at the high end of the range leaves additional room in the calculation to accommodate this potential extra momentum if it is needed.

CHAPTER 10

1. Carl Sifakis, *Encyclopedia of Assassinations* (New York: Facts on File, 1991), p. 109.

NOTES

2. Phillip Kunhardt, "Lincoln's Contested Legacy," *Smithsonian* 39, no. 11 (February 2009): 34.

3. Judith St. George, *Presidents' Lives at Stake* (New York: Holiday House, 1979), pp. 28, 29–30.

4. Will Durant, *The Story of Civilization III, Caesar and Christ*, 11 vols. (New York: Simon & Schuster, 1944), p. 433.

5. Ibid., p. 268.

6. Ibid., p. 284.

7. Ibid., p. 198.

8. Ibid., p. 203.

9. Sifakis, *Encyclopedia of Assassinations*, p. 165.

10. Ibid.

11. St. George, *Presidents' Lives at Stake*, p. 75.

12. Ibid.

13. Sifakis, *Encyclopedia of Assassinations*, p. 165.

14. St. George, *Presidents' Lives at Stake*, pp. 129–30.

15. John S. Craig, "The Guns of Dealey Plaza," http://www.acorn.net/jfkplace/09/fp.back_issues/11th_Issue/guns_dp.html (accessed June 22, 2010), p. 5.

16. Warren Commission, *Report of the President's Commission on the Assassination of President Kennedy* (Washington, DC: GPO, 1964), pp. 546–48.

17. Warren Commission, *Hearings before the President's Commission on the Assassination of President Kennedy* (Washington, DC: GPO, 1964), vol. 24, pp. 829, 831.

18. CIA report 104-40, WC 24, pp. 829, 831.

INDEX

INDEX

INDEX

INDEX

INDEX

INDEX

INDEX

INDEX